# MAKING SENSE OF NUMBER
## Improving Personal Numeracy

Many adults feel that they lack the necessary foundational knowledge in mathematics required to confidently use mathematics in daily life and in their careers. *Making Sense of Number* is a concise introduction to personal and professional numeracy skills, helping readers to become more mathematically competent. It includes relevant content to assist pre-service teachers to improve numeracy for the classroom or to prepare for the LANTITE, as well as support for practising teachers to develop their understanding and skills in numeracy.

*Making Sense of Number* focuses on number sense as a conceptual framework for understanding mathematics, covering foundational areas of mathematics that often cause concern such as multiplication, fractions, ratio, rate and scale. The authors use real-world examples to explain mathematical concepts in an accessible and engaging way. Learning activities throughout the book help readers self-assess their understanding of the mathematical concepts discussed, and answers to activities are included.

Written by authors with over 30 years' experience teaching mathematics at primary, secondary and tertiary levels, *Making Sense of Number* is an essential guide for both pre-service teachers and those looking to improve their understanding of numeracy.

**Annette Hilton** is an Industry Fellow in the School of International Studies and Education at the University of Technology Sydney.

**Geoff Hilton** is an Honorary Research Fellow at the University of Queensland School of Education.

Cambridge University Press acknowledges the Australian Aboriginal and Torres Strait Islander peoples of this nation. We acknowledge the traditional custodians of the lands on which our company is located and where we conduct our business. We pay our respects to ancestors and Elders, past and present. Cambridge University Press is committed to honouring Australian Aboriginal and Torres Strait Islander peoples' unique cultural and spiritual relationships to the land, waters and seas and their rich contribution to society.

# MAKING SENSE OF NUMBER

## Improving Personal Numeracy

ANNETTE HILTON &
GEOFF HILTON

CAMBRIDGE
UNIVERSITY PRESS

# CAMBRIDGE
## UNIVERSITY PRESS

University Printing House, Cambridge CB2 8BS, United Kingdom

One Liberty Plaza, 20th Floor, New York, NY 10006, USA

477 Williamstown Road, Port Melbourne, VIC 3207, Australia

314–321, 3rd Floor, Plot 3, Splendor Forum, Jasola District Centre, New Delhi – 110025, India

103 Penang Road, #05–06/07, Visioncrest Commercial, Singapore 238467

Cambridge University Press is part of the University of Cambridge.

It furthers the University's mission by disseminating knowledge in the pursuit of education, learning and research at the highest international levels of excellence.

www.cambridge.org
Information on this title: www.cambridge.org/9781009009928

© Cambridge University Press 2021

First published 2021

Cover designed by Anne-Marie Reeves
Typeset by Integra Software Services Pvt. Ltd
Printed in China by C & C Offset Printing Co., Ltd, July 2021

*A catalogue record for this publication is available from the British Library*

*A catalogue record for this book is available from the National Library of Australia*

ISBN 978-1-009-00992-8 Paperback
ISBN 978-1-009-03049-6 Paperback and LANTITE bundle

# CONTENTS

## Chapter 7    Fractional thinking    108

## Chapter 8    Ratio, rate and scale    133

# PREFACE

Rather than being a 'how to do' or 'how to teach' mathematics guide, this text specifically aims to assist people who self-identify as needing some additional support in becoming more mathematically confident and competent (although at times we will mention the importance of certain aspects for teachers). The approach taken is to focus on a few foundational concepts that we feel, from our experience, cause common mathematical difficulties for people. The reader is asked to reflect on and self-assess in relation to some core content knowledge and also their personal attitudes to mathematics. From that point, the text is intended for the reader to engage in areas where it is felt assistance is needed. Broadly the focus is to assist the reader to take some steps to improve their number sense. As we progress through the chapters, we will provide learning activities to help clarify ideas and to give readers a chance to apply the ideas. When relevant, the answers to the questions are provided in the Appendix.

## ABOUT NUMBER SENSE

Number sense can be many things to many people, so we have no intention of trying to present the myriad possibilities. We have chosen topics (as reflected in the contents pages) that we feel are at the core of mathematical understanding and being numerate but which we find are the most common areas of concern. Because mathematics can be inherently complex and abstract, understanding foundational elements that develop personal number sense is vital. Whenever possible, authentic scenarios are used to help the reader contextualise the mathematics concepts and give insight into how they are used and why they are so important.

## ABOUT THE AUTHORS

Annette and Geoff each have over 20 years' experience teaching mathematics in secondary and primary schools respectively. They have taught mathematics education courses at tertiary level for the last 15 years. During this time, they have been researching with and working with in-service teachers in the field of mathematics. In more recent years, they have also had considerable experience helping pre-service education students prepare for or re-sit their LANTITE (Literacy and Numeracy Test for Initial Teacher Education), a process during which some common issues have become apparent. These many years of experience across the spectrum of mathematics education have informed the focus of this text.

# Reflection on personal mathematics experiences and abilities

**1**

## LEARNING OBJECTIVES

After reading this chapter, you should be able to:

- describe possible factors that have contributed to your mathematical knowledge and attitude to mathematics
- recall the factors that may contribute to mathematics anxiety
- understand the reasons why we need positive mathematics dispositions
- explain how number sense and a growth mindset can benefit mathematics learners.

# INTRODUCTION

In this chapter we focus on the importance of competence and confidence in mathematics. We will provide opportunities for you to reflect on your own strengths and weaknesses in various areas of mathematics and explore ways to develop skills in areas where you feel you might need to improve. Not only is this important in terms of developing your skills and knowledge around mathematics, but the ability and willingness to engage with learning experiences to strengthen your understanding are key to being a lifelong learner. Rather than provide mathematical learning activities, in this chapter we provide reflection activities to assist you to think about aspects of mathematics that you find challenging.

For the last 15 years we have worked with primary and secondary pre-service teachers in postgraduate and undergraduate education programs as well as primary and secondary in-service teachers. This has given us a privileged perspective on many educators' mathematics learning experiences. It is more common than one might expect to find that many of these people have stories to tell about personal difficulties or issues with mathematics. These challenges take many forms and their origins are often quite complex. Some people report concerns with specific mathematics content areas, some report a general lack of confidence in their mathematics ability, and others have feelings of anxiety aroused by engaging with mathematics. (For example, many people report feeling that working with fractions causes them some problems. This has been noted even by some students who achieve well in mathematics.) Of course, some people have a mixture of these experiences and circumstances. For many people with whom we have worked, their frustration has increased over time as they struggle to effect positive change in their relationship with mathematics. So the pressing question is, how can challenges in one's own interactions with mathematics be addressed and improved?

At this point, we would like to remind readers that this book is not for learning how to teach mathematics to children. It targets adults who feel they need some extra support to become more confident in mathematics. We know that the educators with whom we work have attempted to address their personal mathematics issues (often many times and over a very long time). They sometimes report progress and sometimes frustrations. So we are keenly aware that to improve one's mathematical circumstances is a very personal journey. Our hope is that by using this text to analyse these circumstances, readers may gain some insights, hints, motivations and new ideas to assist them to continue their journey of mathematical self-improvement.

An analogy: What if mathematics were a jigsaw puzzle? People might fall into one of the following categories; they might:

- love jigsaw puzzles and do them easily and often
- enjoy jigsaw puzzles for the challenge they present
- have all the pieces but sometimes struggle to put them together
- always seem to be missing some pieces and become frustrated
- have had a bad experience and prefer not to do jigsaw puzzles.

To improve as a jigsaw 'doer' it would be important to self-assess before beginning to make changes for improvement. Similarly, for mathematics, it is important to determine your personal circumstance and the reasons underlying it before proceeding to take steps to improve. From a mathematics perspective where do you feel you are?

In this chapter we ask that you consider your personal mathematical world. We ask that you be honest with yourself and attempt to clearly identify issues that you may have with mathematical knowledge and **dispositions**. We will discuss dispositions in more detail in Chapter 2, but for now let's say that our dispositions cause us to respond in a particular way. For example, some people might have a number of positive dispositions to learning particular subjects, such as persistence, creativity or curiosity, whereas others might have more negative dispositions, such as fear or disinterest.

> **Disposition** refers to the way in which someone responds to a situation (Claxton, 2014) – for example, with resilience, determination, empathy, confidence.

# WHAT IS MY MATHEMATICAL CIRCUMSTANCE?

As we have discussed in the previous section, we all have our own experiences and personal circumstances related to mathematics learning. In this section we will examine these in more detail. This personal emphasis may seem odd – even paradoxical – in a book designed to help people improve their mathematical skills and knowledge; still, we are going to use a popular mathematical approach to problem solving devised by Polya (1957) to get started on the personal analysis of your mathematical situation. Polya is well known for articulating a set of steps to follow in an attempt to solve a mathematical problem: (1) Understand the problem; (2) Devise a plan; (3) Carry out the plan; (4) Look back at what you've done. We feel that these steps are equally pertinent to solving many problems in life (not just in mathematics, although we will look in depth at this in Chapter 9). While the first step, which involves both identifying and understanding the problem, seems obvious and simple, we humans often tend to do everything but. Some of the classic 'non-identifying' strategies that we are all guilty of from time to time include:

- denying we have a problem (I haven't put on too much weight over the festive season)
- admitting we have a problem but blaming someone else (I've put on a few kilos but if my partner stopped cooking amazing desserts it wouldn't happen)
- hiding the problem (I'll wear a large, sloppy, kilo-hiding sweater for a while).

This is why some (many?) problems in our lives persist; we simply don't honestly identify them so they can be addressed. Life problems, when not identified and addressed, can continue to grow, and cause us great distress (think finances, relationships, fitness and health). The same can happen if we have concerns about mathematics.

## Understanding the problem: a personal perspective

In this section we will begin to consider the possible underlying concerns that you may have in relation to mathematics. These may be related to knowledge, attitudes, or past experiences – or even a combination of these.

---

### Reflection activity 1.1

Write down your own thoughts and feelings about mathematics. You might write about positive or negative aspects of your own learning experiences, your understanding of mathematical concepts, your confidence in your mathematical ability, or how you feel about mathematics (either as a learner or user of maths). Which of these are negative? How might you begin to address these challenges? Which are positive? Is there a way to build on the positive ideas?

---

It is likely that at some stage you experienced either difficulties with mathematical concepts or you developed the perception that you are not mathematical or that mathematics is too difficult for you. It may be that the knowledge gaps developed first and this led to negative attitudes about mathematics or about your mathematical self-perceptions. It might be that you have a belief that even when you try, you will have difficulties and this impacts on your engagement with mathematics (see Wilkie & Sullivan [2018] for a discussion about motivation and mathematics learning). We know that there are strong links between motivation and achievement (Middleton, 2013) and this is something we hope to assist you with in this book. In the next sections we invite you to think about some scenarios and reflect on your own experiences.

### Knowledge gaps

It is amazing that so often people can articulate exactly when they feel their difficulties with mathematics began. They often describe in great detail the situation, the year, the topic, even the teacher they had at the time.

## THE FEELINGS OF OTHERS

Here are a few experiences shared with us over the years. See if you can relate to any of them or perhaps you have your own story (feel free to contact the authors if you want to share).

- I had a period of absence at school and when I returned, I struggled to catch up.
- I changed schools a few times and one time I found myself behind in maths and never caught up.
- When we started learning fractions, the teacher moved forward really quickly and I 'didn't get it'.
- I found it hard to remember all the formulae and rules.
- I felt confident until mathematics became more abstract and letters were introduced.
- One year the teacher and I didn't get on, I just got further and further behind.
- Things were really bad at home and I couldn't concentrate in school.

## Reflection activity 1.2

Do any of these statements resonate with you? Is it possible to identify what you consider to be your main knowledge gaps? Try to write a list that might allow you to identify concepts that you would like to focus on in future chapters of this book.

The experiences described in Scenario 1.1 all resulted in people missing a 'chunk' of mathematics in their school years and we will begin to look at some of the key elements of fundamental mathematical knowledge later in this chapter. These missing chunks can have quite a powerful knock-on effect as the mathematics curriculum builds on itself over the years. The mathematics curriculum is often referred to as a spiral curriculum. The content from one year is used as a foundation for the content of the next as it expands. Therefore, missing a chunk of knowledge from one year can have multiple ramifications for the ensuing years. As these gaps in knowledge grow, another situation often develops: The negative feelings towards mathematics begin to magnify the impact of the initial missing knowledge.

## Dispositional responses

Sometimes past experiences in school or in our personal lives have a profound effect on our attitudes to mathematics or our self-perceptions as mathematics learners. This doesn't happen to all people in this situation but from our experience it happens to many and these negative feelings can be powerful inhibitors to moving forward in mathematics. The group of stories in Scenario 1.2 exemplify how a particular attitude or self-perception may develop.

### Scenario 1.2

## THE SOURCE OF NEGATIVE DISPOSITIONS

In these experiences people described how gaps in their mathematics knowledge impacted their dispositions toward mathematics.

- Missing out on some maths learning meant I got behind and then failing led me to dislike and even fear maths.
- We used to play a maths game where if you didn't get the correct answer you had to stay standing. I hated it and then I hated maths.
- I would be embarrassed when the teacher asked me questions (even when I didn't have my hand up). I would get them wrong and then I would become afraid of being asked another.
- I learned to fear failure in maths, even when I knew the work, and this would often make me so anxious that I couldn't think straight. Things spiralled down from there.
- I'm a high achiever, I expect to do well in everything I do. I couldn't cope when I didn't do so well at maths. It really knocked my confidence and I started questioning my abilities.

### Reflection activity 1.3

Once again, we invite you to reflect on the stories in Scenario 1.2. Are any of them similar to your own experiences? Are your feelings related to identified knowledge gaps or are they more attitudinal or related to your perceptions of yourself as a mathematics learner – or are they more complex?

If you have seen yourself in any of these stories, or you have other experiences that have had a detrimental effect on your mathematics knowledge, ability or confidence then perhaps it can be said that the source of the issue has been identified. Now comes the important time to take some affirming steps. Remember we're not going to deny the problem, blame others or continue to hide the problem. It is natural and easy to blame Teacher X from Year 4 for the way you feel, but that element of the situation cannot continue to be a blockage for your personal progress in mathematics. Our goal through this book is to support you to develop your mathematical knowledge and skills and, as a result, we hope to help you gain confidence and (maybe) even some enjoyment in learning and doing mathematics.

## Been there, done that

Having identified the problem, the second step in Polya's problem-solving strategy is to devise a plan. This may be easier said than done in the case of overcoming problems with mathematics knowledge or anxiety. Again, from our experiences we understand that many people have struggled long and hard to overcome their areas of difficulties with mathematics. It would be insulting just to say, 'Try harder!'.

Using the jigsaw analogy from earlier in this chapter, this text aims, in particular, to help people who have all the mathematics pieces but are not always able to put them together, those who don't have all the pieces and therefore struggle to see the big picture, or those who are sick and tired of trying to find and fit the pieces. This text offers perhaps some different ways of thinking about some core mathematical ideas. We focus on core concepts. We don't cover all possible concepts and we certainly don't cover all concepts in minute detail. As often as possible the mathematics is taken out of the textbook or mathematics classroom and situated in the real world. By doing this, some knowledge gaps can be addressed and the negative mathematics baggage that many readers may carry with them might be assuaged, by engaging in a different but authentic approach. Our intention is to take mathematics into the real world where it becomes numeracy (the focus of the next chapter). By trying to engage with this you can't be accused of doing the same thing over and over and expecting a different outcome (which some might regard as the definition of madness) but you will be able to perhaps view mathematics from a different perspective: an everyday, real-world, practical help to your daily life. Before we focus on the initial steps in devising a plan, let's continue our focus on Polya's first step of understanding the problem.

# UNDERSTANDING THE PROBLEM: A RESEARCH PERSPECTIVE

This section describes the nature of mathematics anxiety, the factors that contribute to it, and its impact on mathematics learners. We have a twofold purpose for this focus. First, it is important to understand that you are not alone and that there are many factors that may have led you to having negative dispositions toward mathematics. Second, and equally importantly, if you are an educator it is essential that you understand how your own students may be feeling and that you do all you can to build their positive dispositions and not impact negatively on your students as a result of your own mathematics experiences. It should also be said that not all people who experience challenges with mathematics have mathematics anxiety and we also acknowledge that you may be using this book simply as a means to improve your mathematical skills and knowledge.

## What is mathematics anxiety?

**Mathematics anxiety** is a feeling of fear or apprehension that can interfere with our ability to use or learn mathematics.

**Mathematics anxiety** is a feeling of fear or apprehension that arises when we are asked to learn mathematics and that can interfere with our ability to use or learn mathematics (Brady & Winn, 2017; Olson & Stoehr, 2019). The relationships between mathematics anxiety and the factors that impact negatively on learners of mathematics are often cyclic in nature; the situation is almost analogous with the chicken and egg scenario: mathematics anxiety can inhibit our ability to learn or understand mathematics just as our lack of understanding of mathematics or past negative experiences can lead to mathematics anxiety. In short, mathematics anxiety can be caused by a number of factors and it in turn can be a significant factor that influences our engagement and persistence in learning or using mathematics.

The key factors that drive mathematics anxiety are related to the relationship between our experiences in mathematics learning and the resultant feelings that we develop. Our past experiences can strongly influence our self-perceptions (Adelson & McCoach, 2011; Bandura, 2001). Negative experiences in mathematics learning during school (especially when young) or poor past performance in mathematics can lead to worry and anxiety about mathematics learning (Eccles & Roesner, 2011). Sometimes we may not be influenced by negative personal experiences but by our own or other people's views that only certain (smart) people are good at mathematics (Brady & Winn, 2017). As mentioned already, mathematics anxiety can also challenge or inhibit our capacity to learn mathematics and thus impact negatively on our chances of achieving in mathematics (see Olson & Stoehr, 2019).

## What is the impact of mathematics anxiety?

When we experience the negative emotions described in the previous section, it is logical that we would not find enjoyment in learning or using mathematics. Indeed, research has shown that when people who are anxious about mathematics are faced with a mathematical situation, their brain responds in the same way as it would if they had detected a threat or were experiencing pain (Lyons & Beilock, 2012). This being the case, it is important that we address the causes of these feelings so that we can start to understand the challenges we may be facing. Mathematics anxiety or the factors that lead to it can result in learners feeling so worried or fearful that their ability to learn and understand is reduced because their working memory is inhibited (Haylock, 2019). A fear of not being able to achieve at mathematics can also develop into learners losing confidence and they may begin avoiding situations involving mathematics or they may disengage from mathematics learning altogether, including in subjects where mathematics is essential for success, such as science (Quinnell, Thompson & LeBard, 2013).

It is fair to say that regardless of whether you experience mathematics anxiety or whether some of the factors that can lead to it are at the core of your concerns about using or learning mathematics, it is important to try to identify some strategies that may assist you to feel more willing to engage with mathematics learning activities and to feel more confident and comfortable with your ability to use and learn mathematics. This is essential for two main reasons. First, it is important for us as individuals to rise to the challenges posed by mathematics. Second, as educators we have to accept our role as numeracy teachers and to recognise that our students need us to be the most numerate we can be.

Always remember that a negative disposition can be a strong barrier for some people, and it takes time and effort to change our feelings. The following anecdote illustrates how the mere thought of mathematics is enough to cause a negative reaction. One of our past teaching colleagues was a brilliant English teacher who was very anxious about mathematics. While supervising a senior mathematics exam during block testing, the teacher was standing next to a student who raised his hand. The teacher rushed across the assembly hall and said, 'you go and help that student, I'm terrible at maths'. It turned out that the student simply wanted to ask for more blank paper!

# WHY IS IT IMPORTANT THAT TEACHERS ADDRESS THEIR MATHEMATICS KNOWLEDGE AND DISPOSITIONS?

Sadly, during their primary school years many children develop beliefs and attitudes about their capability in mathematics as well as whether they enjoy mathematics, and these attitudes and beliefs influence both engagement and achievement in mathematics (Dowker, Bennett & Smith, 2012; Olson & Stoehr, 2019). Those students who have achieved success in mathematics early in school are more likely to continue to engage with mathematics because they have positive attitudes, enjoy mathematics and have high mathematical self-perceptions (Adelson & McCoach, 2011; Goos, Dole & Geiger,

2011). In addition to improved engagement, positive attitudes have a positive influence on students' achievement in mathematics, which makes it imperative for teachers to find ways to enhance their students' attitudes and engagement (Barkatsas, Kasimatis & Gialamas, 2009; Hilton, 2018).

Research has shown that as educators, if we have negative mathematics dispositions or mathematics anxiety we can risk passing these on to our students (Beilock & Willingham, 2014; Burnett & Wichman, 1997). It has been found that mathematics anxious teachers are more likely to use inflexible teaching approaches and emphasise algorithmic thinking over conceptual understanding (e.g. by focusing on procedures rather than the concepts underpinning the mathematics) and memorisation over sense making. According to Olson and Stoehr (2019), teachers who are anxious about mathematics sometimes resort to these ways of teaching because they are concerned about teaching students 'correctly' and fear that they might confuse their students. Such teachers also spend less time helping students with their questions compared to teachers who are not anxious about mathematics (see Ramirez et al., 2018). It is also important to understand that our students are quite aware of what constitutes good mathematics teaching and learning and that they are strongly influenced by effective teaching strategies (Attard, 2011; Reys et al., 2017).

Taken together, these are compelling reasons to ensure that we focus on ways to develop our own mathematics knowledge and our dispositions to mathematics teaching and learning.

# ADDRESSING MATHEMATICS DIFFICULTIES

Earlier in this chapter we described the steps in Polya's problem-solving process. Step 2 involves formulating a plan. In this section we will introduce some ideas that will be central to the approaches used in this book and that we hope will assist readers to begin devising the plan to respond to their own mathematics challenges. Some of these ideas relate to mathematical knowledge and skills while others relate more to your attitudes and beliefs around learning mathematics.

## Number sense

As could be imagined, there are myriad elements to mathematics, many of which incorporate a huge range of detail and levels of complexity. The first decision made when writing this text was which elements of mathematics would be most valuable for people to consider as they try to improve. Broadly, we have called our focus *improving number sense*. **Number sense** is a term that means many things to many people (Berch, 2005) and this can also

**Number sense** is the ability to see, understand and use numbers and to understand how numbers relate and work together.

contribute to people's confusion. Researchers have identified multiple diverse ideas within various definitions of number sense (Tosto et al., 2017). We will not debate the various understandings of number sense and nor will we debate the host of elements that some people wish to include in number sense. Put simply, number sense, as referred to in this text, is the set of knowledges and skills that enable us to see, understand and use numbers; to understand how numbers function and relate to one another; and to understand how numbers work together in different ways to provide us with simple yet powerful personal mathematical and problem-solving abilities (see Berch, 2005; Way, 2011). Number sense in this book encompasses the concepts and strategies dealt with in each of its chapters. The equivalent in literacy would be related to understanding vocabulary, punctuation and grammar to be able to build sentences, paragraphs and longer texts.

People who have a strong number sense are known to be proficient at:

- mental computation
- estimation
- judging the relative size of numbers or quantities
- recognising relationships among numbers and operations
- understanding place value concepts
- moving between different representations of numbers
- problem solving.

As we proceed through the remaining chapters of this book, we will deal in depth with each of these seven elements of number sense.

## Reflection activity 1.4

We will discuss numeracy and continue our discussion of number sense in Chapter 2. As we proceed through the rest of the chapters in the book, we will focus in depth on many of the elements of number sense. At this stage we encourage you to think about the description and elements of number sense presented so far in this section.

Are there any strengths that you feel you have?

In which area(s) would you like to feel more confident?

Before we proceed on our problem-solving journey with Polya, we want to look at the ideas of becoming a lifelong learner and adopting a growth mindset as other strategies to assist us in the endeavour of responding to our mathematical challenges.

# Lifelong learning

As educators we all encourage our students to be lifelong learners. Being able to model this ourselves is an important characteristic of our profession. Teachers who do not learn or who lack the willingness to learn would be a concern. So how do we learn? In particular, how do we learn a topic with which we previously have not had the success we desired? In this latter case there may be scars that have to be healed before progress can be made: this can be very difficult.

There are many stories of people who have overcome barriers to re-engage in an activity; for example, a local surfer was bitten by a shark but six months later returned to the water, though somewhat apprehensively. As difficult as one could imagine this to be, the surfer had one advantage: he loved surfing and as such had intrinsic motivation. People returning to mathematics learning may not have the luxury of a 'lost love of maths' that they are hoping to reignite. So the motivations for 'having another go' at improving personal mathematics abilities have different sources. Educators are responsible for their students' numeracy learning (in Australia, numeracy is a general capability, the teaching of which is part of all teachers' job descriptions) so are motivated to improve for their students. Indeed, this self-improvement is an imperative as a teacher. As well, people know that being confident and competent in day-to-day mathematics has personal benefits in their work and general life (think finances, shopping, budgeting). Ultimately, unlike the surfer bursting with intrinsic motivation, many people re-engage in their personal mathematics learning because they know they should do so (perhaps they have extrinsic motivation from society or their students). As educators we know that intrinsic motivation is usually a more powerful force than extrinsic motivation, so after we have been brave enough to 'identify and understand the problem' we now must make a personal decision about how to address the problem – that is, we need to start to formulate a plan.

Pendergast (2017) proposed some important considerations relating to developing lifelong learning habits. If we are to become more engaged lifelong learners, it is important to think about whether we have these characteristics and habits (and how we can develop them or strengthen them):

- self-directed learning
- self-awareness and metacognition
- critical and creative thinking
- planning
- listening
- memorising
- reasoning and problem-solving skills.

As lifelong learners we need to understand how we learn – when and why do we consider learning important? This involves understanding why we may find it difficult to focus on improving something (especially if we don't like it). It requires the recognition that as a lifelong learner, things may take time and that resilience and persistence may be needed to achieve our learning goals. This can also demand commitment and focused awareness and we need to tune in rather than tune out. Lifelong learners never lose their curiosity – they keep asking questions and have a thirst for developing new skills or knowledge.

Sometimes these things can be easier said than done. Remember that there are people and resources who can help. Technology can be an incredibly useful resource for learning and understanding in any field but there are some particularly great online resources and videos to assist you to learn about, understand and apply mathematical skills and concepts.

## Reflection activity 1.5

So we'd like to be lifelong learners – so what? In this reflection activity we ask you to think about a plan that might help you to embark on the lifelong mathematics learning journey. Analyse your strengths and challenges – in which ways or aspects of your life do you consider yourself to be a lifelong learner? What are the elements that you know you could work on improving? Make a list of new things you plan to try (related to mathematics learning) and start with doable and manageable ideas. As a starting point you might like to think about the past week. Were there ways in which you used (or would have liked to use) mathematics? Look back at the list from Pendergast about the habits of lifelong learners. Are there any that you could strengthen or focus on that might assist you to improve your mathematics knowledge or skills?

## Growth mindset

Related to the belief that we can all improve in our mathematics skills and abilities if we are prepared to commit the time and energy is the idea of a **growth mindset**. According to Grant and Dweck (2003), a growth mindset is the belief that our abilities or intelligence are not fixed and that with sufficient effort they can be grown over time. Imagine if young children never believed that they would be able to learn or do new things! They are constantly asking questions, following their curiosity, wondering.

**Growth mindset** is the belief that abilities or intelligence are not fixed but can be improved over time with sufficient effort.

In contrast to a growth mindset, people with a **fixed mindset** believe that abilities or intelligence are fixed and unchangeable – some people

**Fixed mindset** is the belief that ability or intelligence is fixed and unchangeable.

are just more able to do mathematics than others. Such individuals are more likely to avoid learning opportunities, especially if they expect them to be difficult. They are likely to give up rather than persist and interpret failure as evidence that they will never be able to understand mathematics (Blackwell, Trzesniewski & Dweck, 2007; Ramirez et al., 2018).

We believe that a growth mindset will help you to persist when you are finding learning challenging and it is a good way to help yourself overcome your negative attitudes (bearing in mind that attitudes are very difficult to change and you will need to be kind to yourself and remember that with time and effort you will improve). This brings us to another important aspect of learning. It is still possible to learn even when the content or concepts we're dealing with seem complex and challenging – for many of us, struggling with challenge is what helps us to learn more effectively. Imagine if everything we started out to learn required little or no effort – would it really feel meaningful or interesting? According to Sullivan et al. (2019), learners are more likely to understand and remember the mathematics they have learned if they have engaged in 'appropriately challenging' tasks (p. 32). Of course, the challenge these authors have in mind is one that requires a learner to puzzle over and think deeply about – and this can require us to remain positive and remember to have a growth mindset.

Throughout this book it is our intention to help you see the relevance and usefulness of mathematics and to help you to overcome your mathematical challenges. We hope to show you how the mathematics you may have found challenging in the past can be understood and applied in authentic everyday situations. With time and persistence, you may even grow to enjoy mathematics learning!

## Scenario 1.3

## A CHANGE OF HEART ABOUT NUMERACY

Often in our work with in- and pre-service teachers we meet people who have never developed confidence in themselves as users of mathematics. One such case was a pre-service secondary school teacher who took part in a numeracy across the curriculum subject. By working through the ideas around numeracy and the role of mathematics from the perspective of their teaching areas and in the context of their own personal life, the student came to feel more positive about numeracy. In their own words, 'for the first time in my own education I found myself genuinely engaged in conversations regarding numeracy – without dreading it or feeling stupid'.

The point of Scenario 1.3 is not to congratulate ourselves but to emphasise that if you are prepared to work at it, and if you believe in your skills to improve your number sense and by extension, your numeracy, you have a good chance of achieving those goals. If we didn't believe in your potential, it would seem pointless to write this book!

## Reflection activity 1.6

1. Consider the ideas presented about growth mindset and fixed mindset. Identify a situation in which you feel you learned something effectively. What characterised that learning experience? For example, you might talk about how you felt, why you were motivated or what made you want to persist.
2. Now think about a situation in which you gave up on the learning. What was it that stopped you from persisting? What might have been done to change this situation? How could you approach the situation again with a growth mindset?

# CONCLUSION

In this chapter we have covered quite a lot of ground. We've set the scene for the remaining chapters by asking you to reflect on your own mathematics learning experiences and abilities. While not wanting to overemphasise mathematics anxiety and your past negative experiences, we have also discussed the ways in which these aspects can impact on your willingness and confidence to engage with mathematics. We believe that understanding what lies behind those feelings can assist you to make a plan to move forward. Our goal throughout this book is to provide you with ideas and information to help you to build your mathematics knowledge, skills and confidence – areas that we truly believe that everyone has the potential to improve. The key is to start small and build rather than trying to do too much too soon and becoming overwhelmed. We know that not everyone shares our love of mathematics but as teachers (of any subject) we owe it to our students to be the best teachers we can be. In this chapter we touched on the research findings that illustrate why this is so important. Finally, the chapter looked at ways in which we hope to help you on your mathematics learning journey by developing number sense, your engagement in lifelong learning, and the adoption of a growth mindset. We hope that by working through the materials and activities in this book, you will grow in confidence and that you will develop a positive attitude to learning mathematics so that your students will benefit.

## Personal actions to improve number sense

By following Polya's problem-solving strategy, it is possible to now better *understand the problem* by identifying challenges to your personal optimal number sense and to start to *make a plan* to address the ways you interact with any challenges. The following are suggestions to assist you:

- Using the jigsaw puzzle analogy, determine what you believe best describes your relationship with mathematics/numeracy.

- If you feel there is improvement needed in your mathematics ability, try to decide where your initial difficulties originated.

- Determine contributing factors to your situation.

- Using the seven elements of number sense listed in the chapter, identify your specific areas of strengths or areas for improvement.

- Use the seven characteristics of a lifelong learner (Pendergast, 2017, and see Reflection activity 1.5) and again determine your strengths or areas for improvement.

- Decide if you have a fixed mindset or a growth mindset with regards to your personal mathematics ability.

# 2

# Mathematics and numeracy: the role of number sense

## LEARNING OBJECTIVES

After reading this chapter, you should be able to:

- describe the key elements of numeracy
- apply the numeracy model to identify the key numeracy elements in a given situation
- explain why numeracy is so important in personal and societal terms
- describe what is meant by the term 'number sense' and apply its different elements in simple scenarios.

# INTRODUCTION

We often ask educators, 'why do we teach maths?' The answers usually include 'so that students can get a good job', 'because it assists with everyday life' and 'it helps our society'. What these responses have in common is the use of mathematics in real lives in the real world. So we expect that the mathematics we teach is to be taken out of the mathematics classroom and used to benefit students' lives and society in general. However, the reality is that often, because of knowledge gaps or anxieties in mathematics, some people do not have the confidence or competence to make best use of mathematics. This has a deleterious effect for the individuals involved and a cumulative negative effect on our society. Of course, it can also be envisaged that if one of these people who lack mathematical confidence or competence is an educator, their students may, as a consequence, be disadvantaged. Sadly, we have heard teachers make comments such as 'I'm not good at maths, but it's okay because the students will get it next year' and 'I'm not that good at maths – that's why I teach the young ones'. It is important not to fall into believing that these are legitimate reasons for a teacher not to try to improve their own mathematics skills.

At the end of Chapter 1 we were considering what it might mean to devise a plan (remember this is Step 2 of Polya's problem-solving process, which consists of (1) Understand the problem; (2) Devise a plan; (3) Carry out the plan; (4) Look back at what you've done). This chapter picks up on that focus and begins by discussing the importance of mathematics in the real world. It describes how using mathematics in context (for most of us) is the central purpose of learning mathematics and thus mathematics becomes numeracy. The chapter will examine numeracy, what it means to be numerate, and the importance of numeracy. We will also continue our focus on number sense and its centrality to understanding and applying mathematics. The chapter will focus on the importance of personal number sense and present an overview of some of the central elements of number sense (i.e. place value, number facts, mental computation, relative thinking, problem solving) that are the focus of other chapters of this book. While reading through this chapter, we encourage you to keep your plan formulation (Step 2) in mind and reflect about what you can do to improve your number sense and numeracy. In the words of Liam Neeson's character in the *Star Wars* prequels, 'Your focus becomes your reality.'

# NUMERACY: A REAL-WORLD REASON FOR LEARNING MATHEMATICS

**Numeracy** is the confident application of mathematics in context.

In this section we consider **numeracy** and its relationship with mathematics. We will also describe the crucial role that numeracy plays in the lives of people and society more broadly. Let's begin by focusing on what we mean by the term numeracy.

# What is numeracy?

It often seems to surprise some educators (both pre- and in-service) to discover that numeracy and mathematics are not the same. That said, they are of course, inextricably linked. The simplest way to view their relationship is to consider the idea that when mathematics is applied to an authentic situation it becomes numeracy. Willis and Hogan (2002), as cited by Hogan (2002), defined numeracy as 'intelligent, practical mathematical action in context' (p. 15). Hughes-Hallett (2001) described the difference between mathematics and numeracy, stating that mathematics climbs the 'ladder of abstraction' and rises above context, whereas numeracy 'clings to context' (p. 94). (To understand this distinction, consider the difference between using basic number facts to solve an everyday problem and determining how to solve an algebraic equation).

The last several decades have seen many varied definitions of numeracy, although many have similar ideas within them. In the Australian Curriculum, ACARA makes the connection between numeracy and mathematics by describing numeracy as the knowledge, skills, behaviours and dispositions needed by students to use mathematics in a wide range of situations, including school learning areas and their broader lives (ACARA, n.d.). In considering the many and sometimes differing definitions of numeracy, Goos (2007) developed a numeracy model for the 21st century, shown in Figure 2.1.

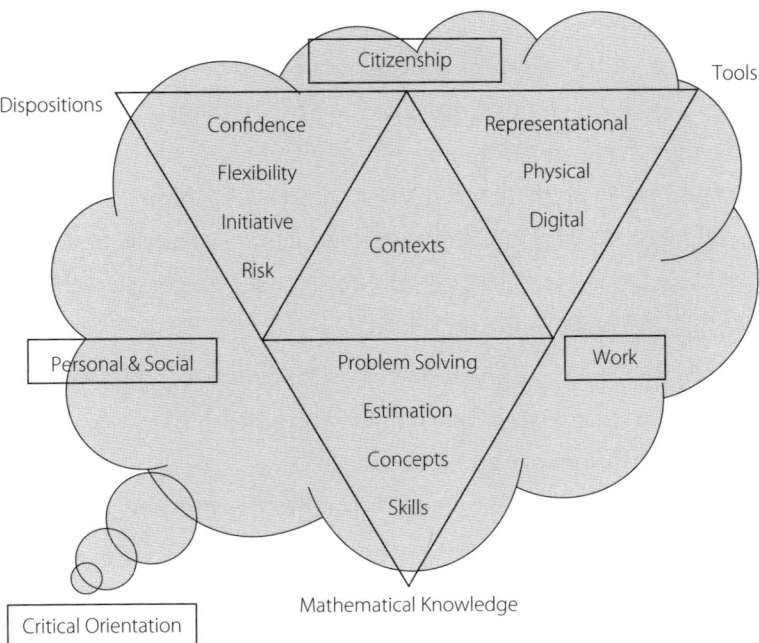

**Figure 2.1** The 21st-century model of numeracy

Source: Goos, Dole & Geiger (2011, p. 132). Reprinted by permission of Springer Nature through the Copyright Clearance Center.

This model represents the key elements of numeracy that are required for a person to be considered numerate. They are:

- mathematical knowledge: mathematical concepts and skills, as well as estimating and problem solving
- the ability to choose and use mathematical tools: digital tools (e.g. spreadsheets, calculators); physical tools (e.g. ruler, tape measure, measuring cup); and representations (e.g. tables, graphs, diagrams)
- dispositions: positive attitudes; confidence; willingness to take risks (e.g. attempt to find an answer), think flexibly (e.g. try a number of different strategies) and persevere
- contexts: real-world and authentic situations, in and beyond the classroom, including work, personal, social and citizenship contexts
- a critical orientation (the ability and willingness to scrutinise, question and challenge numerical information).

## Some everyday examples

The following are examples of everyday activities or situations in which we use numeracy (and probably don't even consciously realise it).

### Example 1. At the petrol station

When I was paying for my petrol last week, the attendant offered me a discount of 8 cents per litre if I purchased any two snack items from the basket on her counter. The cheapest of these items was $3.20, meaning that if I chose two of those items, it would cost me $6.40. I knew that I had pumped around 50 L of petrol so my saving if I chose this offer would be around $4.00. So there was a lot of numeracy happening very quickly; I was under pressure. In addition, I had another important non-numeric question to quickly consider: Did I really want the snacks? While this is a simple example, it shows the use of a number of elements of the numeracy model:

*Mathematical knowledge:* multiplication by two to find the cost of the cheapest snacks ($3.20 × 2 = $6.40); estimation (approximately 50 L × 8 cents per litre = $4.00); comparison of the two amounts

*Tools:* digital (understanding the digital display on the petrol pump), representational (the meanings of mathematical symbols)

*Dispositions:* willingness to estimate the saving and make comparisons between the cost of buying the snacks and not buying them

*Context:* everyday purchasing

*Critical orientation:* willingness to question a situation and make judgements about the 'deal' being offered.

## Example 2. The missing ingredient

While baking muffins I realised that I didn't have any buttermilk. A quick search online revealed that a substitute for buttermilk can be made using regular milk and lemon juice (in the ratio of 1 tablespoon of lemon juice to every cup of milk). I needed 125 mL of buttermilk, so I mixed one half-tablespoon of lemon juice with one half-cup of milk.

In this example the elements of numeracy used were:

*Mathematical knowledge:* conversion of cup measures to fractions and millilitres, knowledge that 1 cup = 250 mL, and understanding and applying ratio

*Tools:* physical tools (measuring cups and spoons)

*Dispositions:* confidence in using the measuring cups and spoons and in converting and using different units of measurement and flexibility in choosing the amounts of milk and lemon juice

*Context:* cooking

In Example 1 a critical orientation was required to be sceptical that the 'deal' was in fact going to be in my favour. In many numeracy situations, critical orientation plays a more pivotal and essential role in supporting our ability to engage and make informed decisions. In Example 2 a critical numeracy orientation was not required – and this is one of the reasons why it is important to experience many mathematical situations in and across contexts. Perhaps it could be said that in Example 2 a person with a critical orientation might decide whether the muffins with the buttermilk substitute tasted different from those made next time with store-bought buttermilk; however, this is not really critical numeracy but rather it is related to taste preference.

As you can see from these examples, not all elements of numeracy may be used in every situation but if we gain enough experiences, all aspects will develop over time.

## Scenario 2.1

### APPLYING THE NUMERACY MODEL

Consider the following scenarios. For each, identify what you consider to be the key elements of numeracy that would be required (e.g. what mathematical knowledge, which tools, what is the context, what essential dispositions are needed, is there a critical orientation involved?).

1. A Year 6 science student investigating how temperature affects the solubility of salt
2. A Year 11 history student learning about the events that led to World War 1
3. A cabinetmaker designing a kitchen
4. A veterinarian making up an anaesthetic injection for a dog
5. A parent planning a budget for the next family holiday

**Figure 2.2**  An interest rate sign at a bank

Think back to our discussion about dispositions and growth mindsets in Chapter 1. There are situations in which we might all feel lacking in our numeracy. The image in Figure 2.2 is an example we found recently. It is a sign that was outside a bank. There are two interest rates displayed on the sign and one is identified as the 'comparison rate'. Neither of us really knew what this meant but as we didn't need a home loan we weren't bothered. We clearly needed more information in order to make sense of the information. However, if someone needed to know, the right disposition would allow them to ask, and to make sense of the information they received.

## The central role of context in numeracy

All of the elements of the Goos model shown in Figure 2.1 are important to numeracy, and this model reinforces the notion that the development of a numerate person requires much more than just a knowledge of mathematics. In fact, as we admitted in our discussion on Figure 2.2, even when an individual has solid mathematical knowledge and skills, there may (and probably will be) situations in which they are not functionally numerate (Hughes-Hallett, 2001). Situated cognition researchers have argued that this is because the knowledge and skills that we learn are often disconnected from context (Brown, Collins & Duguid, 1989). Remember Polya's Step 2? You should be formulating your plan while reading this chapter. What are some everyday contexts that you could deliberately engage in to help improve your own numeracy?

In many ways it could be said that because numeracy is the context-based application of mathematical concepts, being numerate is the true test of our mathematical understanding; however, as we have seen from the definitions of numeracy described already, numeracy encompasses much more. Normally in schools mathematics is learned in class and then the knowledge is used in practical numeracy contexts. The approach taken in this text is, as often as possible, the reverse of this idea: we will examine numeracy contexts and work backwards to examine the underlying mathematical concepts and skills. The key reason behind this approach is that we also want to work on dispositions and other elements of numeracy.

## Why is numeracy so important?

An experience that many of us have had may give some insight into the difficulties of innumeracy. If you have travelled to a country that has an unfamiliar language, it is possible to feel the impact and frustration of being illiterate in that country. That same feeling is an everyday part of life for an innumerate person. An innumerate individual would find it very difficult to function in everyday life. Handling money and finances, estimating, and mental computation, to name a few, would be difficult and cause frustration and difficulties, often detrimental to the person's wellbeing.

It is interesting to think about levels of numeracy. The numeracy demands of particular situations and the ways in which the numeracy demands are placed on a person by society are constantly changing. Consider how rapidly the job market has changed in recent years and the new skills required for success in new, technology-focused occupations. People without numeracy skills can literally be left behind as the job market changes. It is also important to think about our attitudes to literacy and numeracy. Many schools and government policies over recent years have focused on literacy to a greater extent than numeracy. There may be multiple reasons for this. Parents sometimes value literacy more than numeracy as an essential life skill and indeed we have heard parents say, 'We are a literacy family – we don't care so much about numeracy.' It may also be that illiteracy is perceived as less socially acceptable than innumeracy (Sellars, 2018). This can be reinforced when we hear others say, 'Not everyone is good at maths – I was never very good at maths either.'

Research has shown that low levels of numeracy can be a greater hindrance than low levels of literacy to employment opportunities. This can have subsequent impacts on self-esteem, health prospects, and participation in society (Council of Australian Governments, 2008; Geiger, Forgasz & Goos, 2015). Many people find the idea that innumeracy can be more detrimental than illiteracy to be counterintuitive, but it is difficult to think of any job that would not require some use of mathematical skill (numeracy). To think a little more deeply about this, it is revealing to consider the opposite situation. What would it be like to be an innumerate individual? What would it be like to have a critical mass of innumerate citizens? How would such members of our community actively engage with their world? How could they make sense of the numbers they hear from politicians?

If a society fails to ensure its citizens are literate and numerate then the ramifications for that society can be dire. In this modern, globalised, competitive world, some societies or countries are finding it difficult to keep up. The consequences often include economic hardships. Interestingly, most of these countries turn to education to address the situation. Governments realise that having a literate and numerate society is a priority (sadly there are exceptions – for example, those who withhold education from certain people within their population, such as girls and women or minority groups).

In a democracy it is important for the citizenry to be well informed so they can make decisions about their choice of government and that government's policies. This is not a new idea; over 20 years ago Steen proposed numeracy as the key to understanding our 'data drenched society' (1998, p. 8). Many decades earlier Dewey argued that for democracy to succeed individuals must be able to think for themselves, make independent judgements, and distinguish good information from bad (Dewey, 1931–32). In short, citizens need to be able to interpret what they are being told by politicians and through the media, which requires them to be critically numerate and literate. In these days of dubious news sources, these skills are even more important. One doesn't have to look far around the world to see examples of where this criticality is beginning to erode. Thus we as individuals have an important role to play in developing personal numeracy for both our own benefit and that of society.

The question remains: What level of numeracy does a person need? This is the 'how long is a piece of string' conundrum. The simple answer for numeracy is as much as possible. No two people have the same numeracy skills. The demands of our lives (social, work, personal, political) usually define what is necessary, and these can change quickly (think about how time management changes when we have children, start a new course, or get a job for the first time). As we have said already, much of the numeracy with which we engage on a daily basis may not be recognised as such. Quite often numeracy happens as part of our everyday thinking. For example, consider your morning routine. The time you get up may be determined by the commitments you have on that day, or before you leave the house; whether you eat a leisurely breakfast or rush out the door may depend on the time you get up! You may have to estimate whether you have enough fuel to drive to work and back – and whether you have enough cash or credit left to pay for any required fuel – or how long that trip to work will take in peak-hour traffic. These are all examples of everyday incidental numeracy where we are probably not even conscious of using mathematics. Most situations such as these rely on our basic number sense and our experiences. Note also that in such cases we are unlikely to make formal mathematical calculations; rather, our number sense allows us to make estimations and approximations to inform our decisions. In the next section, we will focus on number sense in more detail.

# THE KEY IDEAS OF NUMBER SENSE

As we described in Chapter 1, number sense is fundamental to our ability to understand and apply mathematics in any context. It is especially important in situations that are new or that require us to be able to solve problems. Remember also from Chapter 1 that we said that people who have a strong number sense are known to be proficient at:

- mental computation
- estimation
- judging the relative size of numbers or quantities
- recognising relationships among numbers and operations
- understanding place value concepts
- moving between different representations of numbers
- problem solving.

In this section we will continue our focus on number sense and its centrality to understanding and applying mathematics. Here we will use a series of scenarios and activities to introduce some of the central elements of number sense and the skills that someone with number sense has. It is actually difficult and sometimes not possible to talk about each of these in isolation of the others. For example, problem solving relies on all the other elements. These key ideas will be expanded upon in the remaining chapters of this book. Our goal in this section is not to examine any of these in detail but to provide you with opportunities to begin to consider where you feel confident. If in the following scenarios and learning activities we ask you questions about which you are not sure, do not be concerned. Try your best and remember that the purpose of the remaining chapters is to assist you to develop the skills and knowledge you need. Perhaps you could use the learning activities as a personal diagnostic tool to find out more about where you have a solid understanding and where you could plan to improve. Remember the discussion about lifelong learning and a growth mindset from Chapter 1!

## Mental computation

**Mental computation** involves the use of number facts and elements of number sense to be able to calculate without the aid of tools, such as pen and paper or a calculator. The first step clearly is knowing the number facts that allow us to perform addition, subtraction, multiplication and division. This allows us to use efficient strategies to perform required calculations.

### Scenario 2.2

### USING NUMBERS IN EVERYDAY LIFE

1. In recent years, the families in my neighbourhood have organised Hallowe'en activities involving households who are happy to participate. I am stocking up on chocolates because the children in my neighbourhood will come trick-or-treating. I have 7 packets of chocolates, each of which contains 15 individually wrapped chocolates. I wonder whether I have enough chocolates.
2. My favourite netball team is playing against their arch-rivals, and at half-time my team is behind and the score is 35 to 19. I'm thinking about how far behind my team is …

### Learning activity 2.2

Consider the situations in Scenario 2.2. Without using pen and paper or a calculator, calculate the answer to each. What operation did you use in each? How did you know what to do? Describe the strategies you used to help you.

*(Answers in Appendix.)*

## Estimation

Estimating is a set of skills that are useful in many situations and in many occupations. Estimation involves more than simple guesswork because it requires us to use our number sense to arrive at an answer. It also allows us to make judgements about the reasonableness of calculated answers. Estimation is often used in situations where we want to find an approximate answer, but it is also useful in spatial situations, such as judging dimensions or distances between objects.

## Scenario 2.3

## CAN I ESTIMATE?

1. I am painting a rectangular wall of my garage and the dimensions are 6.8 m × 3.2 m. The label on the paint can says that 1 L of paint covers 16 m² of wall. I need to decide whether I have enough in my 2 L paint can to paint the wall. I'm also wondering what I will need if I need to paint it more than once.
2. I am driving to Townsville and would like to break my journey with an overnight stay around halfway to Townsville.

   Soon after I start my journey, I pass the road sign shown in Figure 2.3. I need to decide where I will stop for the night.

| Brisbane | 166 |
| Gympie | 332 |
| Rockhampton | 798 |
| Townsville | 1 516 |
| Cairns | 1 861 |

**Figure 2.3**   Road distances

## Learning activity 2.3

For each of the situations in Scenario 2.3, use estimation to find a solution – you do not need to make detailed calculations and, again, this should be possible in your head. In each case are there any assumptions you need to make?

*(Answers in Appendix.)*

## Judging magnitudes

The ability to judge and understand the absolute and relative magnitudes (or sizes) of numbers or quantities is important for a number of reasons. It allows us to comprehend individual numbers and quantities and to make comparisons between or among quantities. This skill is also useful for considering whether your answer to a mathematical problem is reasonable and feasible.

---

## Learning activity 2.4

1.  Choose the largest number in each of the following:
    a.  104, 140, 400
    b.  0.104, 0.14, 0.099
    c.  23.034, 23.34, 23.304
2.  It's 2020 and I have to decide how many people I can have in my café according to the government rules about social distancing. My available floor space is 65 m². The minimum space for each customer is 4 m². I calculate that I can have 65 ÷ 4 = 16.25 people in my café. Does this answer seem reasonable? What do I need to do?
3.  My best ever time for running 100 m is 13 seconds. I estimate that I should be able to run a 10 km race in 1 300 seconds. Is my logic correct? Why or why not?

*(Answers in Appendix.)*

---

If you found parts b or c in Question 1 of Learning activity 2.4 challenging, do not be overly concerned – we will spend quite some time in the next chapter discussing place value and the interpretation of decimal places.

## Place value

As mentioned above, we will talk at length about place value in Chapter 3. In short, place value is a key element of our number system whereby the position of a digit within a number gives it its value. People who have number sense are able to use place value concepts easily and accurately to deal with numbers. Place value understanding is one of those pervasive elements of number sense that helps us in using and applying some of the other elements of number sense. (In the first question in Learning activity 2.4, you would have used your understanding of place value to help you find the answer.) Place value understanding is needed to understand and compare numbers, to name numbers and to calculate accurately.

## Numerical relationships

This term covers a range of mathematical concepts, including operations (e.g. addition, subtraction, multiplication, division); proportional reasoning (relative thinking, **multiplicative thinking**, fractional thinking, rate, ratio, scale); and properties of numbers (e.g. factors, multiples, commutativity, associativity, multiplicative and additive identity, distributivity). These ideas will be further developed as they are the focus of future chapters of this book where we will expand on their definitions and provide more examples.

> **Multiplicative thinking** is the ability to work flexibly with numbers, concepts, representations and strategies of multiplication and division (Siemon et al., 2011).

## Representational fluency

As with numerical relationships, representational fluency encompasses a number of aspects. It involves the ability to understand, interpret, use and create representations for numerous purposes as well as the ability to transform and move between different representations. Some simple examples include being able to create or interpret mathematical diagrams, tables and graphs. Another skill associated with representational fluency is understanding equivalent representations – for example, 0.1, 10%, one-tenth.

# EVERYDAY APPLICATIONS OF NUMBER

In Figure 2.4 there are a number of everyday signs and images. Think about the information contained in each.

**Figure 2.4**   Everyday examples of mathematical representations

## Learning activity 2.7

For each of the images in Figure 2.4, describe the mathematical information they contain and suggest a context in which they would be used (or be essential).

## Problem solving

Again, this term covers a range of possibilities. We have already introduced Polya to you and we will expand on problem types and the strategies and processes involved in problem solving in the last chapter of this book. People who are good at problem solving are able to use their number sense and draw on its many elements to identify what the problem is and which information they will need to solve it. They are able to understand which operations they will need to use and in what order. They are also able to check the accuracy of their answer and judge its reasonableness. In Scenario 2.5 we have presented two everyday situations that we could use number sense to engage with.

## Scenario 2.5

### WHAT CAN I AFFORD?

1. I pass my favourite shoe store and notice that the shoes I have been wanting (which were priced at $150) are on sale. The sign says that they have been reduced by 20%. I'm still not sure if the sale price will fit with my budget. I need to know how much the shoes will now cost.
2. I am interested in learning to surf. I am not very coordinated so I estimate that I will need at least two lessons. Being a backpacker on a lengthy holiday, I am on a tight budget so would like to choose the option that will get me the best value for money. Using the prices listed on the surf shop wall (shown in Figure 2.5), I'm keen to compare the deals to find out which deal will allow me to pay the least amount per lesson.

| | |
|---|---|
| 1/2 DAY LEARN TO SURF ADVENTURE | $69 |
| 2 DAY PROGRESSIVE LESSON | $120 |
| 3 DAY ULTIMATE WAVE RIDER | $165 |
| 4 DAY EPIC SURF EXPERIENCE | $200 |
| 5 DAY AMPED SURF TOUR | $240 |
| 5 DAY DELUXE INCLUDES A PRIVATE ONE ON ONE LESSON | $350 |

**PRIVATE LESSONS**

**SINGLE SURFER**
2 HOUR LESSON

| | |
|---|---|
| ONE ADULT | $160 |
| ONE CHILD | $140 |

**TWO SURFERS**
2 HOUR LESSON

| | |
|---|---|
| TWO ADULTS | $260 |
| TWO CHILDREN | $240 |

**Figure 2.5**  Surfing lessons price list

## Learning activity 2.8

For each of the situations in Scenario 2.5, think about how you might solve the problem. Think about your mathematical thinking – which operations (+, –, ×, ÷) did you use? How did you know what to do? What were you unsure about? What else would you like to know?

*(Answers in Appendix.)*

# CONCLUSION

In this chapter we have focused on two important aspects of mathematics in our everyday lives: numeracy and number sense. These ideas are at the heart of this book and are essential not only for our personal and social lives but for our capacity to work and engage actively in and with our civic and natural worlds. Numeracy incorporates more than mathematical knowledge – it is the use of that knowledge in context and includes the ability to use mathematical tools as well as the positive dispositions to persevere and be a flexible risk taker. We use our number sense in order to function in situations requiring numeracy. It is difficult to define number sense succinctly, but a person who has a strong number sense has the ability to reason and solve problems using estimation, mental computation, place value, representations, and their knowledge and understanding of numbers' magnitudes and meanings, and the relationships between them.

## Personal actions to improve number sense

By following Polya's problem-solving strategy, it is possible to now better *understand the problem* by identifying challenges to your personal optimal number sense and to start to *make a plan* to address the ways you interact with any challenges. The following are suggestions to assist you:

- Using the Goos model of 21st-century numeracy, superimpose your understanding of your personal numeracy to identify strengths or areas for improvement.

- For a short time (a day or two), take note of or photograph all the situations in which you encounter a numeracy moment in your life.

- Analyse your interactions with these encounters.

  - Were you confident?

  - Did you engage?

  - Did you avoid anything?

  - If so, what was the problem?

3

# The Hindu-Arabic number system

## LEARNING OBJECTIVES

After reading this chapter, you should be able to:

- explain the origins of the Hindu-Arabic number system
- explain the importance of place value and give examples
- explain the importance of Base 10 and give examples
- explain the importance of zero and give examples
- identify some 'naughty' mathematics shortcuts that you should never have been taught.

# INTRODUCTION

When we work with teachers and pre-service education students, we find that very few are able to identify our number system as the Hindu-Arabic number system. The reasons for this are a mystery, other than it seems to be a piece of knowledge that has become lost in time or deemed unnecessary. However, knowing the name of our number system hints at its history, and knowing the history and reasons for its wide acceptance provides core understandings for developing number sense.

In this chapter the history of the Hindu-Arabic number system is discussed and the main features of the system and how they form the foundation for number sense are explored. There is also a section about some 'naughty things' that you may have been taught in the past, but should not have, because they undermine your understanding of the number system and number sense.

# A BRIEF HISTORY OF THE HINDU-ARABIC NUMBER SYSTEM

The Hindu-Arabic number system is used in many countries around the world today. It has a fascinating history and developed over time many centuries ago. The following is an abbreviated version of this history.

The Hindu-Arabic system, as the name suggests, is a melding of ideas from two parts of the world. The digits 1 to 9 were acquired by Middle Eastern traders from their dealings with India. The use of zero was later incorporated (also believed to originate from India). The concept of the position of a digit giving it value through its place being held by zero was a simple but powerful idea. For example, 700 attains its place value by the 7 being held in the third column (the hundreds column in **Base 10**) by two zero placeholders. We will return to these important features later in this chapter.

A **Base 10** number system uses 10 digits to represent all numbers (De Klerk, 2014).

It wasn't long before the real power of this system was revealed in a book written by al-Khwarizmi in Baghdad about the year 875, which showed how the Hindu-Arabic system could be used for calculations. The book has numerous title translations, one of which is 'The Science and Restoring and Balancing' (as in balancing equations). The author, al-Khwarizmi, lent his name to 'algorithm' and one of the Persian words in the book title, 'al-Jabr' became today's word 'algebra'. The ease of calculation afforded by the Hindu-Arabic system was an efficient and effective boon for traders and trading. It also paved the way for extraordinary advancements in mathematics and science. This system slowly spread via traders across the Mediterranean Sea and into Europe, eventually supplanting the initially dominant Roman system.

Some historical accounts of the development of this system vary and many details are no doubt lost in time. However, as earlier stated, to understand the story of the Hindu-Arabic system is to understand the foundations of number sense. Some searches online will reveal much more detail and many more intriguing stories about this amazing system and its history. For an interesting video about the mathematician al-Khwarizmi, you could try https://www.aljazeera.com then search for 'Al-Khwarizmi: The Father of Algebra'.

The development and wide acceptance and use of the Hindu-Arabic number system in many ways parallels much of the Mediterranean, Middle Eastern and subcontinental history of the Middle Ages. Countries and cultures developed many of their own systems – for example, systems of government, language, measurement and laws. As countries began to interact more, mainly through trade (and sometimes wars), a problem would often arise: whose laws, whose language or whose measurement system would be used? Number systems were no different. Whenever trade is at the heart of interactions, efficiency and effectiveness are always central tenets because these lead to higher profits. So when number systems used by different traders overlapped, some systems were slowly sidelined (e.g. the Roman number system) and some were merged, as in the Hindu-Arabic system, and thus grew in importance.

While the Hindu-Arabic number system seems very simple – it has only 10 symbols (i.e. the digits 0 to 9 – Base 10) and any number can be formed by using these symbols in varying positions – it took centuries to be fully developed and many more centuries before it was accepted. To understand the reasons for the rise of the Hindu-Arabic number system and the decline of the Roman is to take a very important step to developing personal number sense.

The Roman number system with which you may be familiar is still used today but mainly for things like chapter numbers in books and numerals on clock or watch faces, as shown in Figure 3.1. It is interesting to note that old clocks (such as in medieval cities and towns in Europe) have Roman numerals. Why might that be?

**Figure 3.1**  Examples of clock faces with Roman numerals

The Roman number system was very useful for counting but was not very easy to use in calculations because it does not have place value, does not use Base 10 and has no zero. The Romans did, however, give us the Latin origins for mathematical words, such as calculator and calculation. Roman numerals were so ineffective for calculating that when a calculation was needed, people often worked things out using small stones in a sand tray. The small stones were called *calcs*, which is the Latin origin of the words used today. To understand why place value helps with calculations, consider the Hindu-Arabic and Roman versions of the simple addition problem shown in Figure 3.2. Which seems to you to be more logical and would be easier to complete?

```
      3  4                    X  X  X  I  V
   +                       +
         8                     V  I  I  I
      _____                   _____
      4  2                     X  L  I  I
```

**Figure 3.2**   Hindu-Arabic and Roman addition

Note how the lack of place value would make calculations in Roman notation complex. Imagine if we tried to multiply XXXIV by VIII. This is one of the reasons why the Hindu-Arabic system replaced the Roman system – because it allows for mathematical procedures to be performed  (remember this all happened before anyone had calculators!).

So, the Hindu-Arabic number system has particular features that make it widely accepted and extremely functional. Over the centuries, as the world has globalised, many other common systems have developed, such as the metric system for measurement or dollars and cents for currencies. These systems also owe their existence to the Hindu-Arabic number system, because of its very specific features.

As we have already mentioned, the Hindu-Arabic number system has three main features. Understanding these features is central to developing number sense. It is like many things in life in which understanding how something works can increase its utility and be very advantageous to the user. The three features of the Hindu-Arabic number system are place value, Base 10 and zero. In the remaining sections of this chapter, we will look in detail at each of the three defining features of the Hindu-Arabic number system that give it such power and underpin our development of number sense.

# FEATURE NUMBER ONE: PLACE VALUE

Place value is a feature that allows us to create numbers by virtue of the location of the digits 0 to 9. It also helps us to name numbers, especially when they are large. Place value is a core concept at the heart of a great deal of mathematics. From our experience of many years teaching school students and university students, the lack of a clear understanding of place value is often at the heart of mathematical difficulties being experienced. This observation is also supported by research, especially when it comes to understanding decimal place values. For example, a study by Moloney and Stacey (1997) revealed that around 25% of Year 10 students could not compare decimals with a high degree of accuracy.

Before looking at decimal place value we will introduce the idea of place value for whole numbers.

## Placement of digits in place value columns

Place value is the concept of a digit's position in a number giving it value. Consider the numbers 385 and 263, also shown in Table 3.1. Both numbers have a 3 in them, but the 3s have different values because of where they are placed. In the first number the 3 is in the hundreds column and its value is 300, whereas the 3 in the second number is in the ones column and has a value of only 3. (As a counterexample, the Roman number system does not have place value, which is why it is so hard to use for calculations, as seen in Figure 3.2 in the previous section.)

**Table 3.1** The numbers 385 and 263 in place value columns

| Hundreds | Tens | Ones |
| --- | --- | --- |
| 3 | 8 | 5 |
| 2 | 6 | 3 |

In the early years of schooling it is usual for children to learn about the columns of place value. This starts with children counting up to 10 and experiencing the digit 1 moving left to the tens column and the digit 0 holding a place in the ones column to form the number 10. To achieve this, a vertical format (as opposed to writing the numbers in the horizontal form 1, 2, 3, 4, …) is used so that the movement to the tens column can easily be seen. Table 3.2 illustrates this. Teachers of young children sometimes refer to the columns as 'houses'. They will speak of the ones house or the tens house. As it takes a number of years for students to progress through the 'houses' (the hundreds column is usually not introduced in the first year or two of schooling), children's understanding of place value gradually expands and develops. As children begin using numbers in calculations, the teacher usually emphasises the importance of maintaining place value. It is for this reason that old-fashioned neatness and setting out remain very important as they allow for accuracy in maintaining place value. Again, students we see who have mathematical difficulties sometimes seem to have a disorganised approach to setting out their work and, as observers, we have great difficulty following what they are trying to do. Mathematics is the science of structure, order and relations evolved from counting, measuring, and describing shapes (Gray, 2019). Keeping mathematical working in a structured format (i.e. acknowledging place value as shown in Table 3.2) improves the chances of accuracy and in the long term enhances number sense.

**Table 3.2** A simple place value chart

| Hundreds | Tens | Ones |
|---|---|---|
| | 1 | 0 |
| | 1 | 1 |
| | 1 | 2 |
| | | 3 |
| | | 4 |
| | | 5 |
| | | 6 |
| | | 7 |
| | | 8 |
| | | 9 |

Looking again at Table 3.2, you can see that it represents Base 10 (because it has 10 digits in the ones column and then moves left to the tens column) and that the only digits used are 0–9. Once children count to 9, a 1 must be placed in the tens column to show 10, 11, 12, and so on. On this chart the number 28, for example, would be represented by 2 tens and 8 ones. Place value is also adhered to in many real-world situations. Consider the way that the distances on the highway sign are represented, as shown in Figure 3.3. The numbers are listed with attention to place value.

**Figure 3.3** Road distance sign: an example of place value in everyday signage

# Learning activity 3.1

In this activity we will play a game using a calculator. The game is called *Calculator Crash*. The aim is to use your knowledge of place value to reduce a given number to 0 by subtracting. You will need one die to designate the digit you are allowed to use and a calculator to play the game. The steps involved are:

1. Put your chosen five-digit number into your calculator (only use digits 1 to 6 as per digits on a die) – for example, 35 243.
2. Roll the die and note the value on top of the die – the value on the following die is 3.

3. If that value matches a digit in your number, you are allowed to remove the corresponding digit (and only that digit) from your calculator number by subtraction using the value you rolled on the die (as shown in the example below).
4. Repeat Steps 2 and 3 until you reach 0 on your calculator.

Let's work through an example. Imagine that your chosen calculator number is 35 243. Next, the die is rolled and it shows a 5. You notice that your number has a 5 in the thousands column, so you can subtract 5 000, which gives a result of 30 243:

$$
\begin{array}{r r}
35 & 243 \\
-\phantom{0}5 & 000 \\
\hline
30 & 243 \\
\end{array}
$$

Your next roll gives you a 2. You have a 2 in the hundreds column, so you subtract 200 from 30 243, which results in 30 043.

Can you see that with each subtraction you are creating zero place holders where the previous digits have been? This process continues until you have a final result of 0. Let's say that you score a 3 on your next roll. You have two 3s in your number: a 3 in the ones column and a 3 in the ten thousands column. You can choose either to subtract 3 or to subtract 30 000. If you subtract 30 000, your result will be 43. If you then rolled a 4, you would subtract 40, resulting in 3. You would then need to roll another 3 to subtract your final digit and obtain a score of zero.

Your turn: try playing the game with a five-digit number, 23 651; a six-digit number, 651 643; or create your own, even larger number.

Learning activity 3.1 can also be played as a game where pairs of players take turns to see who can get to zero first. The number of digits entered in the calculator can be adjusted to suit the players' level of understanding.

The large numbers used in Learning activity 3.1 can be difficult to say. In the following section we will look at another valuable advantage of place value – it helps us to name numbers.

## Naming numbers

Place value is also the key to another very important foundational skill: saying large numbers. Being able to accurately say numbers enhances the ability to make numerical comparisons. Numbers are written and read in groups of three: for example, 598 621 (note the space between the groups of three digits). The first group of three digits on the right (621) are in the hundreds, tens and ones columns, so six hundred and twenty-one. The second group of three digits (598) repeats the pattern but this time the digits are in the hundreds, tens and ones of the thousands columns, so five hundred and ninety-eight thousand. To read the number completely the repetition of the hundreds, tens and ones can be heard: five hundred and ninety-eight thousand, six hundred and twenty-one. Figure 3.4 shows the groupings up to the billions group and another example.

Place value groupings

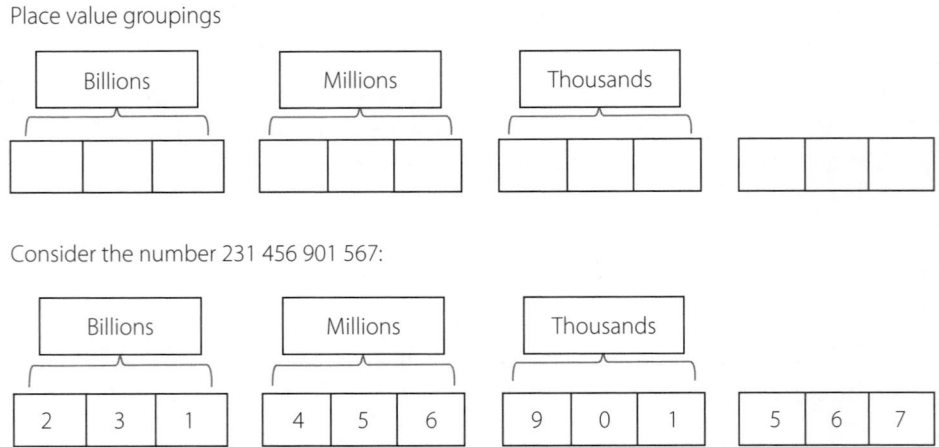

Consider the number 231 456 901 567:

We would say the number as two hundred and thirty-one billion, four hundred and fifty-six million, nine hundred and one thousand, five hundred and sixty-seven.

**Figure 3.4**   Place-value groupings

As you can see, the pattern is continued into the next group of three digits, which are the hundreds, tens and ones of millions and so on.

## Learning activity 3.2

In this activity we want you to practise naming large numbers. While this may seem tedious, it is a good chance to practise identifying the place values and groups of three. Write down how would you say:

1. 32 509 354
2. 1 254 098 465
3. 908 196 740 031

*(Answers in Appendix.)*

Leaving a space between each set of three digits is sometimes substituted with a comma. This is an old practice but can still be found today. Figure 3.5 shows an example of this from a real estate window. In both listings the prices have a comma to delineate the groups of three. The example on the left shows the price range with the two numbers aligned one above the other in the same place value (which allows easier comparison) whereas the prices on the right are not place value aligned.

| | |
|---|---|
| **6 BED** | **4 BED** |
| **4 BATH** | **3 BATH** |
| **6 CAR** | **6 CAR** |
| **36 ACRES** | **12 ACRES** |
| **$2,199,000 -** | **$1,900,000 -** |
| **$2,299,000** | **$2,100,000** |

**Figure 3.5** Examples of price formats from a real estate agent's sales window

Again, there are some possible disruptors to number sense around large numbers. For example, in Australia, mobile phone numbers are arranged into a four-digit prefix and then groups of three digits, such as 0493 609 734, while land-line numbers are grouped in four digits – for example, 6685 2961. The grouping of these numbers is more about convenience of remembering the number, and of course the place values are not given, only the digit names. These numbers are not intended to represent quantities – they are just **nominal** numbers. Other examples of nominal numbers are postcodes, driving licence numbers and passport numbers.

**Nominal** numbers are used for naming or identifying.

Having explored place value and the naming of whole numbers, we will now turn our attention to place value and the naming of decimal numbers.

## Decimal place value

Place value continues into the decimals but with an interesting twist. The number 32.6 has digits in the tens, ones and tenths columns. What many young learners ask is where is the 'oneths' column? Quite naturally they feel the need for symmetry around the decimal point. They believe it should be tens, ones, decimal point, oneths, tenths. However, if the decimal point is seen as signifying the end of the whole numbers and sits in the ones column, then everything is symmetrical around the ones column. This is illustrated in Figure 3.6. Of course, Figure 3.6 only shows a small number of place values; however, the place value columns continue indefinitely in both directions.

The place value columns are symmetrical about the ones column

| Hundreds | Tens | Ones . | tenths | hundredths |

**Figure 3.6**   The place value columns are symmetrical around the ones column (and thus, the decimal point)

It is also worth noting that in this globalised world, sometimes a European decimal is encountered, which is a comma instead of a point: for example, €1,80 and €2,50, as shown in Figure 3.7. This photo is from a French farmers' market. Note the use of the comma instead of a decimal point in the prices.

**Figure 3.7**   An example of the use of a comma instead of a decimal point

## Naming decimal numbers

An unusual habit that many of us have (and sadly many of us were taught) is the way we say decimal numbers. Most of us would correctly say 3 740 as three thousand, seven hundred and forty. Notice that in this correct method the digit is stated followed by the place value, even though four tens has been shortened to forty. This correct method is important because it reinforces the place value every time that we say a number and thus it strengthens our number sense. However, something strange happens when we move to numbers with decimals. How would you say 345.62? The usual replies we hear might be some of the following:

- three hundred and forty-five point six two
- three hundred and forty-five decimal six two
- three hundred and forty-five point sixty-two

Each of these replies shows an interesting phenomenon: they articulate the place value for the whole numbers but fail to do so for the decimals. These replies are more indicative of common usage in our lives; we would hear things like this on the television news for instance. The correct way of saying 345.62 is three hundred and forty-five and *sixty-two hundredths*. Notice that now the decimal is attributed its place value name of hundredths. While this might be seen as pedantic, if a person never learns the correct method then they are again denied important foundational support for their number sense. This is especially important for teachers to emphasise with children who have just begun learning about decimals so that they develop a strong understanding of the decimal place values. Interestingly, this abandonment of place value names for decimals tends to coincide with the introduction of this more complex fractional concept; so just when things get more difficult (moving from whole numbers to decimals) the place value support is withdrawn. Also notice that the decimal point is called 'and' when saying the number correctly.

## Learning activity 3.3

Write down the way that you would say the following numbers:

1. 34.5
2. 487.23
3. 1 305.004

*(Answers in Appendix.)*

# FEATURE NUMBER TWO: BASE 10

The second main feature of the Hindu-Arabic number system is that it is a Base 10 system. This feature is strongly related to place value in that the placement of a numeral in a particular position gives its value, and this value is multiplied by 10 when moving to the left from one column to the next column. Conversely, it is divided by 10 when moving right from column to column. See Table 3.3.

**Table 3.3**  Place value columns showing different representations

| Thousands | Hundreds | Tens | Ones . | tenths | hundredths |
|-----------|----------|------|--------|--------|------------|
| 1000s | 100s | 10s | 1s | $\frac{1}{10}$ s | $\frac{1}{100}$ s |

Recall that the first column of the whole numbers is the ones column, the second column to the left is the tens column ($10 \times 1$) and the third column to the left is the hundreds column ($10 \times 10$). Consider the numbers 457 and 538. Both numbers have a 5 (the 5 in the first number represents 5 tens and the second number has 5 hundreds). The second 5 (500) is 10 times greater than the first 5 (50) because the number system is Base 10, and each digit has place value.

Now let's consider decimal numbers. As we move from the ones column into the decimal places, the first decimal place value is the tenths column ($1 \div 10$ or $\frac{1}{10}$) and the second decimal place column is the hundredths column ($\frac{1}{10} \div 10$ or $\frac{1}{100}$). Think about the numbers 2.53 and 1.15. Both numbers have a 5 (in the first number, the 5 is in the tenths column and in the second, the 5 is in the hundredths column). The second 5 is 10 times smaller than the 5 in the first number.

There are other bases we use quite often in our lives. With time, for example, we could count in days: 4 days, 5 days, 6 days, 1 week. This is an example of Base 7. If we add 4 days to 5 days, we could answer with 9 days if we were thinking in terms of Base 10, but if we were thinking in terms of Base 7 (in weeks), the answer would be 1 week 2 days. The relationship between seconds, minutes and hours is Base 60, so we move from 58 seconds, 59 seconds, to 1 minute … 58 minutes, 59 minutes to 1 hour.

You may have heard of binary notation (e.g. 1100111). This system is used in electronics. The term refers to numbers that are Base 2 (and hence it uses only two digits: 0 and 1). Each time we move left from one column to another, the place value is multiplied by 2. You can learn more about this system online if you are interested (try https://www.mathsisfun .com/binary-number-system.html).

# FEATURE NUMBER THREE: ZERO

The third feature of the Hindu-Arabic number system is that it has a zero (some number systems do not!). This seems an innocuous idea, but it is what makes the system work and it has immense power. The main use of zero is to give value to other digits in a number. For instance, which amount of money would you prefer: $6 or $60? Assuming you said $60, you can see the number system at work: the 6 is in the tens column (10 times greater than the ones column – Base 10) and the position of the 6 is being maintained by the zero (place value and zero).

So, zero is a strange concept because it is often perceived as having no value, and yet it is a number. Its power in terms of place value is that it gives other digits value as in the example above. Zero's role is to signify that there is no amount in a particular column. Imagine the amazing mind that first thought of such a concept. We tend to take zero for granted but it really is the cornerstone of our number system and our number sense. The idea of zero has a chequered history. For many people in the Middle Ages used to working with the Roman system (no zero), the idea of zero was frightening. It was declared by some to be a devil's number; after all, how can something be multiplied by zero (a concept that many a young child has pondered and a concept that many teachers have struggled to explain)? Some online exploring will reveal much more detail about the history of zero.

The three features of our Hindu-Arabic number system – place value, Base 10 and zero – make it an extremely efficient and powerful number system. However, it has sometimes been inadvertently undermined, leading to a loss of understanding and thus hindering our number sense.

# SOME THINGS WE SHOULD NOT HAVE LEARNED

There are many things in life for which, over time, we develop shortcuts. Shortcuts are deemed handy as they get us to our destination in a shorter time or by a more direct route. However, by following shortcuts we sometimes miss out on something, such as beautiful scenery, quality time or an understanding of place. In mathematics we are sometimes taught shortcuts that lead to correct solutions, but they too have disadvantages. You may have had a teacher who said, 'just do this, it will get you the right answer, even if you don't know why'. The problem is that we have come to rely on these shortcuts when we work mathematically and, sadly, core understandings become lost to us. This in turn limits our development of number sense because we no longer use foundational concepts of our number system. Here are just two examples of things that we perhaps should not have been taught! (You may be able to recall others from your early mathematics learning.)

# Multiplying and dividing by powers of 10

Consider multiplying a whole number by 10: for example, 37 × 10. You may have been told that the answer may be obtained by 'just adding a zero', meaning to place a zero after the 37, making it 370. While this works, it denies the understanding of place value, Base 10 and the function of zero (Hindu-Arabic number system). To fully understand what happens in this simple multiplication, one must: (a) realise that multiplying by 10 means because our number system is Base 10, the answer will be found by sliding the number one place to the left from the tens and ones columns to the hundreds and tens columns; (b) invoke the concept of place value (37 is made 10 times bigger being placed in its new columns, so it becomes 370); and (c) maintain the new position of the 3 and the 7 by using a 0 to hold the ones place to make 370. Look at Table 3.4. In it you can see the digits 3 and 7 have each moved one place to the left – and that the 0 has been inserted as a placeholder in the ones column to complete the resulting number.

**Table 3.4**   Multiplying 37 by 10

| Hundreds | Tens | Ones |
|---|---|---|
|  | 3 | 7 |
| 3 | 7 | 0 |

Thus it can be seen that the shortcut (of just adding a zero) certainly saves a lot of explanation (and time) but denies a great deal of understanding of our number system. One can imagine if a child were taught the shortcut and not the thorough explanation relating to the number system: their development of number sense would be hindered. There is another problem related to this shortcut, which is the complete misrepresentation of 'adding zero'. Consider the answer to 37 + 0. Hopefully you got 37 as your answer. In other words, just adding 0 changes nothing! Sometimes, in an attempt to simplify or shorten things, opportunities to develop number sense are lessened or lost, and even worse, misunderstandings or misconceptions can develop.

Even when children have quite strong understanding of numbers and appear to be achieving well in mathematics, shortcuts such as 'just add zero' can hinder their understanding. Research by Downton, Russo and Hopkins (2019) found that when highly capable mathematics students in upper primary school were asked to multiply numbers together, they used this method without a clear understanding of the concepts underpinning it. (For example, when asked to multiply 23 by 200, the response might be, 'you double 23 and then add two zeroes'.) Downton and colleagues (2019) referred to this as the 'magical zero approach' (p. 239).

When we are dealing with decimals, this shortcut becomes even more treacherous. Consider what happens when we are asked to multiply 3.2 by 10. The shortcut 'just add a zero' would result in 3.20. Can you see why this shortcut is not correct in this example (3.2 = 3.20)? Clearly, we must keep the decimal point where it is and move the digits to the left. Hopefully you noted that the correct answer for 3.2 × 10 is 32.

# Learning activity 3.4

*Calculator aiding number sense example:* Your calculator can be useful for seeing the digits move to the left when multiplying by a power of 10.

Put the number 572 into your calculator. Now enter × 10 and then =. You should see the number slide to the left to become 5 720, with the zero now holding a place in the ones column. This gives a correct vision of what happens when multiplying by 10. Clear the screen and re-enter 572. Now enter + 0 and then =. You should see that nothing happened at all. Hopefully it can be seen that this shortcut of 'just add 0' does not help our number sense.

While on the topic of zero; this is the only name that should be used for the digit 0. Other common terms such as the name of the letter O (which we pronounce 'oh') and naught (meaning nothing) should not be used, especially when teaching young children, even though these are common usage and accepted in some countries. This saves confusion and again links clearly into the Hindu-Arab number system. One could imagine the possibility of number sense confusion for a young child who hears the letter O is also a number (listen to people giving their phone numbers and note how often they say the letter O instead of zero). This highlights another problem that can arise in developing number sense: what a child hears outside the classroom can sometimes confuse, contradict or disrupt what they are learning in the classroom. These disruptions can persist into adulthood.

Do you recall how you may have been told to solve this: 45.9 divided by 10? If you responded by shifting the decimal point one place to the left to make 4.59, then again you have been taught a shortcut that works but is not correct in terms of the Hindu-Arabic number system. It does acknowledge the idea of Base 10 and place value but has a fatal flaw in that the decimal point does not move. When multiplying or dividing by a power of 10, the digits move to the left or right of the decimal point. So, the solution to 45.9 divided by 10 is to slide the number one place to the right (leaving the decimal place unmoved) to obtain the answer of 4.59. This is shown in Table 3.5.

**Table 3.5** Dividing 45.9 by 10

| Tens | Ones . | tenths | hundredths |
|------|--------|--------|------------|
| 4 | 5 . | 9 | |
| | 4 . | 5 | 9 |

# CONCLUSION

This chapter focused on the basics of our number system and its attributes. Remember that it is called the Hindu-Arabic number system and its characterising features are place value, Base 10 and zero. While these elements may have seemed very simple and obvious, in this chapter there were opportunities to re-emphasise some extremely important foundational concepts of number sense. If you are keen to improve your personal number sense, then taking the time to know the history of the Hindu-Arabic number system is important. Also taking the time to reconsider how numbers and calculations are completed by honouring the requirements of the Hindu-Arabic number system may result in improved number sense. Perhaps avoiding mathematical shortcuts (or at least thinking them through and understanding why they have come about) will help refocus on the underlying beauty and power of our number system.

## Personal actions to improve number sense

By following Polya's problem-solving strategy, it is possible to now *carry out the plan* to improve your personal number sense. The following are suggestions to assist you:

- Search online to learn more detail about the history of the Hindu-Arabic number system.

- Analyse your habits and thoughts about the use of our number system. Do you recognise the importance of place value, Base 10 and zero?

- Change your speech habits when naming numbers:

  - to only use the word zero when referring to the number '0'

  - to correctly express decimals.

- Monitor your calculator use. Are you using it when you don't need to?

# 4 : Number facts

## LEARNING OBJECTIVES

After reading this chapter, you should be able to:

- explain the importance of knowing number facts
- understand how to determine which number facts could be improved
- understand and explain how the number facts work with our Hindu-Arabic number system to form a cornerstone of number sense
- appreciate how and why number facts are the basis for most of our day-to-day calculations.

# INTRODUCTION

In this chapter we will explore the learning of number facts. Number facts are the core addition, subtraction, multiplication and division facts on which most of the day-to-day numeracy tasks in which we engage are based. Examples of these facts are $5 + 6 = 11$; $13 - 8 = 5$; $7 \times 3 = 21$; $24 \div 6 = 4$. It is important that we know number facts so that we are able to engage in everyday tasks with confidence. As well as focusing on the importance of number facts, in this chapter we will reflect on the ways in which people learn number facts – including your own experiences – and will suggest some strategies and tools to use if you think this is an area that you would like to further develop.

As we have mentioned previously, number facts are an essential building block of our number sense. Some of the chapters in this text focus on core number system features such as Base 10, place value, and zero (Chapter 3), while others focus on application of number sense in areas such as estimation and mental computation (Chapter 5) and problem solving (Chapter 9). However, number facts are purely and simply a necessary foundation upon which other areas of number sense rely. So, there's no getting around not knowing number facts; in the next section of this chapter we will explore some ideas about how you might have learned number facts before looking at ways to aid improvement in your personal experience with number facts.

# LEARNING THE NUMBER FACTS

Many people we assist do not have automaticity with their number facts. In relation to number facts, **automaticity** is the automatic recall of number facts (mainly addition, subtraction, multiplication and division). There are many things in life for which we develop automaticity: Experienced drivers change gear in a car without thinking most of the time, whereas learner drivers are very conscious of every gear change; musicians concentrate very hard on notes, chords and timing when learning a song but eventually they will play the song with automaticity. Another way of thinking of automaticity is fluency. If we are fluent in a language, we don't consciously think about each word, we simply speak – we have automaticity with the language, its structure, vocabulary and grammar.

> **Automaticity** refers to something that occurs automatically or without conscious thought or mental effort.

Think of some things in your life that you do with automaticity. Reflect on:

- how you achieved automaticity in that skill or behaviour
- whether there are things in life that you are not able to do with automaticity but would like to
- whether there are some things that you do that don't require automaticity.

Not having automaticity for number facts is a disadvantage for an individual's numeracy. Automaticity allows a person to focus on the problem at hand and not be hindered by trying to figure out a number fact. For example, three friends eat lunch at a café, and they decide to split the bill equally. If the total cost of lunch is $27.90, how much does each person owe? Someone who knows their number facts can answer $9.30 without using a calculator or pen and paper, by dividing $27 by 3 and 90 cents by 3. Someone else might use their number sense to estimate more than $9 but less than $10.

Clearly, lack of automaticity hinders the development of number sense in areas such as estimation, mental computation, and problem solving. Of course, the ramifications of this are exemplified by the behaviour of many people we meet through their avoidance of engaging in mathematical situations. As demonstrated in Goos' model for numeracy in the 21st century (shown in Figure 2.1 and discussed in Chapter 2), a lack of these basic number skills or the positive dispositions (e.g. confidence) to use them can impact on a person's work, personal and social lives, and also their role as a critical citizen. So, if you do not know your number facts with automaticity, this chapter will help you develop this skill.

The way in which we learned number facts may vary considerably from one person to another. Usually they are learned through a progression in our schooling years – that is, certain facts are learned in one year and others in the next and so on. Teaching methods have changed considerably over recent decades. Some approaches are effective, and some are less so. Some teaching methods we have seen appear to be unnecessarily complex, while others attempt to use several different methods in close progression in an attempt to cater for diverse learners. For some students, confusion about some of these methods has resulted in deficits in accuracy, speed and confidence. It must also be said that not all confusing or complex methods are from the recent past. The example in Scenario 4.1 illustrates an old method for multiplying two numbers. (It comes from a book written in 1967, in which it was described in the 'Mathematics in History' chapter.) Incidentally, it is not a method we recommend as a first step to learning multiplication facts! We are not suggesting that you use this method – and do not be concerned if it is difficult to follow – that is the point of including it. We are illustrating the fact that some methods are complicated and can cause more stress for students who already find mathematics challenging.

Scenario 4.1

## AN EXAMPLE OF A COMPLICATED METHOD

Suppose you wish to calculate $9 \times 5$. First write the numbers underneath each other:

9

5

Then for each number, write its difference from 10 (i.e. the answer when you subtract it from 10) next to it:

| Number | Difference from 10 |
|---|---|
| 9 | 1 |
| 5 | 5 |

The digit in the **ones** column of the answer is found by multiplying the numbers in the difference column (i.e. $1 \times 5 = \textbf{5}$).

Draw crossed lines as shown:

The digit in the **tens** column of the answer = the diagonal difference (either of them), which in this case is **4** (i.e. $9 - 5$ or $5 - 1$).

This means that the final answer to $9 \times 5 = 45$.

**Source:** Parkes et al., 1967.

If you didn't follow what went on in this example, *we are on your side*! This example is not intended to present a useful means of calculating multiplication number facts. It aims to illustrate why some people become unnecessarily confused when presented with new or unusual ways of doing things. Hopefully, it also serves as a rationale for learning your number facts until you achieve automaticity! If you enjoy a challenge you might like to work out how this strategy arrives at the correct answer. Try some other number facts to check.

If you know that you still experience gaps in your number facts (even with automaticity), try to remember how you were taught and perhaps identify what caused your concerns – for example, quick progression, overly complex methods, loss of confidence or interest, or gaps caused by varying teachers, absence, or movement between schools.

In the next sections we will look at addition and multiplication facts and start to think about ways to strengthen your familiarity with them.

## Addition facts

The usual sequence for learning number facts is to engage in the addition facts first and then in conjunction with subtraction facts. Because our number system only has 10 digits (0 to 9), we only need to learn addition facts up to 20 (10 + 10) and the related subtraction facts. Most people report being reasonably confident with addition and subtraction number facts. The helpful hint here is that of number fact families. Take, for example, the number fact 4 + 7 = 11 – the three numbers involved form a set of related or derived number facts (often referred to as a fact family). Knowing this can give the learner a quick insight into the related or derived number facts of 7 + 4 = 11, 11 – 7 = 4 and 11 – 4 = 7. So learning one fact really gives us knowledge of four. (If you are a teacher or parent of children in the early years, the Kids Academy video 'Fact Family Triangles – Addition and Subtraction Cartoon', available on YouTube, might be useful for helping them understand simple addition and subtraction fact families: https://www.youtube.com/watch?v=9IhZDEffyTk).

---

### Learning activity 4.1

For each of the following number facts, list the related number facts. The first has been completed for you.

a.  4 + 8 = 12          related facts: 8 + 4 = 12; 12 – 4 = 8; 12 – 8 = 4
b.  7 + 8 = 15
c.  11 – 8 = 3
d.  17 – 9 = 8

*(Answers in Appendix.)*

---

## Multiplication facts

Multiplication facts (and thus divisions) are usually found to be more difficult by learners because they have answers to 100 (10 × 10) as opposed to the addition facts to 20. They are also a little harder to visualise; we might be able to see in our mind's eye 9 + 3 but may have difficulty seeing 9 × 3. Once the concepts of multiplication and division have been experienced, there is a progression through the school years as various number facts are learned. Before we continue, let's revisit some terms related to multiplication number facts you will have heard before, namely factors and multiples. In the example 9 × 3 = 27, the numbers 9 and 3 are called **factors** of 27, and 27 is a **multiple** of 3 and 9. (For the purpose of learning number facts, a multiple is the answer to a multiplication number fact.)

Thinking about addition and multiplication for a moment, there is usually a teaching progression for the transition from addition to multiplication. A method called **skip counting,** where you say all the answers to a set of multiplication facts – for example, 4, 8, 12, 16, 20, and so on – is often used as an initial step in the development of multiplication concepts. Do you remember trying to skip count the 5 times number facts to 100 in one breath? Skip counting can involve more than this, such as beginning from any number; however, in terms of learning number facts, skip counting involves starting at zero and counting up in multiples of a given number. Another step along the way is **repeated addition**: for example, 4 + 4 + 4 is the same as 3 lots of 4 (i.e. 3 × 4).

**Factors** are pairs of numbers that multiply together to produce another number.

A **multiple** is a number that results from the multiplication of two numbers.

**Skip counting** is counting by the same amount (other than 1) each time. For example, 2, 4, 6 … is skip counting in 2s.

**Repeated addition** refers to adding the same number continuously as an alternative to multiplication, the same number – e.g. 5 + 5 + 5 …

## Learning activity 4.2

Try skip counting and using multiplication facts to count the blades on the wind turbines in Figure 4.1.

**Figure 4.1**   Wind turbines, each with three blades

The transition to multiplication can take quite a lot of time. We have had students in junior high school who still use repeated addition to calculate a multiplication question, with often very lengthy working out down the side of their page. The reasons for this are no doubt very complex but it seems, because of the heavy emphasis on addition (and subtraction) in the early years of schooling, these become the 'go to' strategies because they are in the students' 'happy place'. However, this over-reliance on repeated addition ultimately means that the multiplication number facts are not embedded in the students' learning. So, if you find that this is a strategy that you use then it is a signal to you that further focus on multiplication and division is needed. This is particularly important as we will find out later in this book when we explore the importance of additive and multiplicative thinking. In fact, inability to work with multiplication facts with automaticity is one of the key reasons students experience difficulties later in mathematics because it hinders their capacity to use multiplicative thinking in areas such as ratio, proportion and algebra (Goos et al., 2017).

## Learning activity 4.3

Figure 4.2 is not a three-headed giraffe! How might we determine the number of legs we cannot see?

**Figure 4.2** Three giraffes

> What would this calculation look like:
>
> a. for skip counting?
> b. for repeated addition?
> c. as a multiplication fact?
>
> *(Answers in Appendix.)*

Some other common strategies we see in classrooms that are used by students who are struggling with number facts include the use of physical assistance, such as number fact grids and calculators. These have the effect of allowing the child to have some success (especially when moving from number facts to calculations); however, at some stage it is important to help the child attempt to reduce reliance on these aids. If you feel that these aids are still of benefit for you when using number facts, it is important to try to reduce your reliance on them, especially if you are to be able to progress with mental computation in Chapter 5.

Before we continue our focus on multiplication (and division), let's take a moment to discuss ways of learning number facts.

## Approaches to learning number facts

When you were in school, it would have been unlikely that your teacher simply put the number facts on the board and asked you to learn them. Instead, your teacher probably used a number of strategies to help you. These may have included hands-on materials, chanting the facts and looking at number properties and relationships. You would have had (hopefully) many experiences, especially with hands-on materials, where you could learn the concepts of what it means to add and subtract or multiply and divide. Having had these experiences, at some point it would have been advantageous to learn the facts 'off by heart' so they could become the building blocks of further mathematical adventures in primary school. Interestingly, somewhere in the relatively recent educational past, the idea of 'rote' learning was often set aside, and the teacher would have been seen as 'old fashioned' if they had engaged their class in this way. The downfall of rote learning comes if and when this is the only form of teaching and learning of number facts. As we emphasise throughout this book, it is always important to develop conceptual understanding first by engaging learners with a range of experiences (such as using physical materials and play-based activities); however, at some point there is no harm doing some 'rote' learning to enhance speed and accuracy (automaticity). The argument that children get bored with rote learning again only really holds true if it is the only form of learning to which they are exposed. We have had many classes who have sung and danced their way through the number facts on a daily basis, even into the lower secondary years, and the best description of this occurrence was 'fun'.

So if you still lack automaticity with some number facts then perhaps some form of daily practice (just for a while) may be in order. You might like to look at YouTube for some catchy tables songs. For example, Class Dynamix, '7 times table fun song (multiply 7)', https://www.youtube.com/watch?v=wNxMhf9x-cE, focuses on the 7 times number facts, and Laugh Along and Learn, '7 Times Table Song (Cover of Happy by Pharrell Williams) Easy Learn Skip Count', https://www.youtube.com/watch?v=5XT3vxohtBg, skip counts in 7s, which helps you learn the multiples of 7). Alternatively, get an app on your phone – perhaps a number fact quiz app – or put your earphones on and listen to some number fact singalongs. There are many examples of such songs and apps online. An example of an app is Maths Rockx (see https://www.mathsrockx.com). Just as an aside, when searching online for examples, we found the skip counting in 7s song referred to earlier was so infectious that it was still playing our heads a day later! This is an example of why such methods can be effective.

## Learning activity 4.4

If you feel that you have not yet developed automaticity with the multiplication facts, explore the online resources that aim to assist with this knowledge. Choose one that you think will be motivating and spend some time learning a set of number facts you are not so confident with (e.g. the 7 or 9 times tables).

## Why learn number facts?

Because of the structure of our number system, knowing the basic number facts unlocks the door to calculations for any sized number and it gives us the capacity to engage in more complex mathematical reasoning more fluently and confidently. Knowing the facts is an enormous gift of efficiency and power. A musician who practises in private will be aiming for the time when they take their skills into a public performance (even if it is just with friends at home); a sportsperson practises (trains) so they are as ready as possible for the next competition. The same can be said of learners of number facts; they are preparing for the next mathematical performance (numeracy), be it in class or in authentic real-life situations. Unlike music or sport, participation is not voluntary, being innumerate is not an option, so the basics have to be practised either as a child or now as an adult. If you still feel that you do not have automatic recall of all the number facts (+, −, ÷, ×), then an audit is required. In the next section we will explore ways to audit your knowledge of the number facts and look at some strategies for building your knowledge.

# PROPERTIES TO ASSIST THE LEARNING OF NUMBER FACTS

There are many properties of numbers and operations that are important in developing our number sense. They can also be useful for helping us to learn number facts. In this section we will look at some important properties of numbers that can help us to cut down on the number of facts we need to learn.

## Working with zero and one

Zero and one are numbers that have properties that allow us to quickly add some number facts to our list of those we know with automaticity. Let's look at these in turn and see what happens when we multiply and add with them.

### Working with zero

We know that when we *add 0* to a number or *subtract 0* from a number, the value of that number remains unchanged. The number zero is known as the **additive identity**, so called because adding it to a number or subtracting it from a number does not change the identity of that number. This means that you automatically know all your zero addition and subtraction facts (e.g. $7 + 0 = 7$, $0 + 9 = 9$, $8 − 0 = 8$).

> Zero is called the **additive identity** because adding or subtracting 0 leaves a number's value unchanged.

In contrast, when we *multiply any number by 0*, the result is 0. That means you already know the zero multiplication facts too. Examples of such facts are $9 \times 0 = 0$, $0 \times 5 = 0$. (Note that division by zero is not defined.)

### Working with one

One is a number with special properties – some of which are beyond the scope of this book. An important property that helps you know another set of number facts is that when a number is multiplied or divided by one, its value remains the same. So, $9 \times 1 = 9$; $9 \div 1 = 9$. This property is the reason that 1 is known as the **multiplicative identity**. This important property means that you already know all the multiplication (and related division) facts for the number 1.

> One (1) is known as the **multiplicative identity** because multiplying or dividing a number by 1 leaves its value unchanged.

# Commutativity

**Commutativity** is a simple but powerful property of numbers for addition and multiplication. The idea of commuting is usually related to going to work and then coming home again at the end of the day. So, we go one way to work and then we come back the same way (usually). When we apply the idea of commutativity to addition and multiplication, it means that the result is the same no matter which way we do the calculation – this allows us to 'do' a number fact one way and come back the other; for example, $5 + 6 = 6 + 5$ or $8 \times 7 = 7 \times 8$. This is very handy to remember when learning addition and multiplication number facts as it halves the number of facts we have to remember. Note that commutativity does not work for subtraction or division. For example, $9 - 3 \neq 3 - 9$ and $28 \div 4 \neq 4 \div 28$. Figure 4.3 shows a real-life example illustrating the commutativity of multiplication.

**Figure 4.3**   Egg trays arranged to show $2 \times 3$ and $3 \times 2$

Even though commutativity doesn't work for division and subtraction, remember the fact family relationship among numbers to help with learning subtraction and division facts. For example, we know that $4 \times 7 = 28$ and $7 \times 4 = 28$, so we also know that $28 \div 4 = 7$ and $28 \div 7 = 4$. The following learning activity focuses on these related or derived facts.

## Learning activity 4.5

Choose all correct responses to these questions and in each case think of an explanation for your choice.

# AUDITING NUMBER FACT KNOWLEDGE

There are many number properties that help us to develop strong number sense. We will cover more of them in later chapters of this book. Now that we have looked at some of the important properties that contribute to number fact knowledge, we will use addition and multiplication grids as audit tools for number facts. We focus on these as the related subtraction and divisions are implicit within them. Remember the discussion earlier in this chapter about knowledge of one number fact helping us to know related or derived number facts (or fact families). The purpose of number fact grids is twofold: first, they are a ready reckoner if you feel the need to do some regular practice; and second, they are great to colour-code facts of which you have automatic recall and those that you don't. Remember that for many adults these facts are well known but it just might be time for a confidence boost or there is a need for a little more speed and accuracy. It might be that you have become a little rusty because you rely on your calculator – or you just might not use some number facts very often. Number fact grids are also available online if you would like to download some and colour-code known or unknown number facts.

## The addition grid

The addition grid for the number facts up to 10 + 10 is shown in Table 4.1.

**Table 4.1**   The addition grid

| + | 0 | 1 | 2 | 3 | 4 | 5 | 6 | 7 | 8 | 9 | 10 |
|---|---|---|---|---|---|---|---|---|---|---|----|
| 0 | 0 | 1 | 2 | 3 | 4 | 5 | 6 | 7 | 8 | 9 | 10 |
| 1 | 1 | 2 | 3 | 4 | 5 | 6 | 7 | 8 | 9 | 10 | 11 |
| 2 | 2 | 3 | 4 | 5 | 6 | 7 | 8 | 9 | 10 | 11 | 12 |
| 3 | 3 | 4 | 5 | 6 | 7 | 8 | 9 | 10 | 11 | 12 | 13 |
| 4 | 4 | 5 | 6 | 7 | 8 | 9 | 10 | 11 | 12 | 13 | 14 |
| 5 | 5 | 6 | 7 | 8 | 9 | 10 | 11 | 12 | 13 | 14 | 15 |
| 6 | 6 | 7 | 8 | 9 | 10 | 11 | 12 | 13 | 14 | 15 | 16 |
| 7 | 7 | 8 | 9 | 10 | 11 | 12 | 13 | 14 | 15 | 16 | 17 |
| 8 | 8 | 9 | 10 | 11 | 12 | 13 | 14 | 15 | 16 | 17 | 18 |
| 9 | 9 | 10 | 11 | 12 | 13 | 14 | 15 | 16 | 17 | 18 | 19 |
| 10 | 10 | 11 | 12 | 13 | 14 | 15 | 16 | 17 | 18 | 19 | 20 |

When looking at this grid it is important to think about the related or derived number facts referred to in the previous sections of this chapter. For example, when the number fact of $5 + 8 = 13$ is embedded, you instantly have the related number facts of $8 + 5 = 13$, $13 - 8 = 5$ and $13 - 5 = 8$. While these related facts are obvious, it is still important to say/learn them (don't just learn the $5 + 8 = 13$ fact) so that automaticity is developed in all four related facts.

## Learning activity 4.6

Using an addition grid either self-audit or ask a friend to ask you some random facts. Colour-code the ones you are sure of and then you can focus on polishing the ones that need attention. Don't forget to practise the related facts too.

## The multiplication grid

The multiplication grid for the number facts up to $10 \times 10$ is shown in Table 4.2. Again, with this grid if you feel that you do not have automaticity for all the facts then it is important to identify the ones for which you do (maybe download a grid). Most people find that the total amount of number facts that they actually need to practise is quite limited. Most are fine with the 0, 1, 2, 5 and 10 multiplication number facts. This is because the zero number facts are all answered with zero, the answers to the ones number facts are the number by which 1 is multiplied, the twos number facts are just doubles, the fives number facts all have an answer ending in 0 or 5, and the tens number facts have the answer in the next column with a zero placeholder (our number system at work again). So, at this point half the number facts on the multiplication grid have already been covered. The 3, 4, 6, 7, 8 and 9 number facts are usually the focus for learners who need to improve automaticity. But even here, because of the commutative law of multiplication, the number of facts to be learned is halved; learning $6 \times 7$ is the same as $7 \times 6$.

**Table 4.2**   The multiplication grid

| × | 0 | 1 | 2 | 3 | 4 | 5 | 6 | 7 | 8 | 9 | 10 |
|---|---|---|---|---|---|---|---|---|---|---|---|
| **0** | 0 | 0 | 0 | 0 | 0 | 0 | 0 | 0 | 0 | 0 | 0 |
| **1** | 0 | 1 | 2 | 3 | 4 | 5 | 6 | 7 | 8 | 9 | 10 |
| **2** | 0 | 2 | 4 | 6 | 8 | 10 | 12 | 14 | 16 | 18 | 20 |
| **3** | 0 | 3 | 6 | 9 | 12 | 15 | 18 | 21 | 24 | 27 | 30 |
| **4** | 0 | 4 | 8 | 12 | 16 | 20 | 24 | 28 | 32 | 36 | 40 |
| **5** | 0 | 5 | 10 | 15 | 20 | 25 | 30 | 35 | 40 | 45 | 50 |
| **6** | 0 | 6 | 12 | 18 | 24 | 30 | 36 | 42 | 48 | 54 | 60 |
| **7** | 0 | 7 | 14 | 21 | 28 | 35 | 42 | 49 | 56 | 63 | 70 |
| **8** | 0 | 8 | 16 | 24 | 32 | 40 | 48 | 56 | 64 | 72 | 80 |
| **9** | 0 | 9 | 18 | 27 | 36 | 45 | 54 | 63 | 72 | 81 | 90 |
| **10** | 0 | 10 | 20 | 30 | 40 | 50 | 60 | 70 | 80 | 90 | 100 |

## Learning activity 4.7

Colour-code the multiplication grid to identify the facts you need to practise (and yes, don't be afraid to rote learn them, sing them, listen to them or do aerobics to them in preparation for your next numeracy performance). Using your multiplication grid either self-audit or ask a friend to ask you some random facts. Again, don't forget to learn the related facts. If you find rote learning challenging or if you would like to explore other ways of developing automaticity with number facts, you may wish to read the article by Jo Boaler about fluency (see https://www.youcubed.org/evidence/fluency-without-fear/). Bear in mind that the strategies mentioned in the article are more suited for use in a classroom than for individual adults.

In this section we have looked at auditing your number fact knowledge. In our surroundings it is often possible to identify examples of situations in which multiplication and division can be visualised. In the following section we will look at a common example: arrays. Arrays can be useful for visualising and practising multiplication facts.

# ARRAYS FOR LEARNING NUMBER FACTS

Earlier we described the way that young students experience the precursors of multiplication through skip counting and repeated addition before learning the multiplication facts. Learning about skip counting and repeated addition and subsequent initial multiplication can be enhanced through hands-on experiences with

> An **array** is a rectangular arrangement of rows and columns containing quantities, symbols or objects.

**arrays**. Arrays can be drawn, created, and searched for in the real world. It is perhaps a little sad to admit that we have been known to deliberately go looking for arrays in real life (array safaris if you like). We have included some of our own photos of everyday arrays in Figure 4.4.

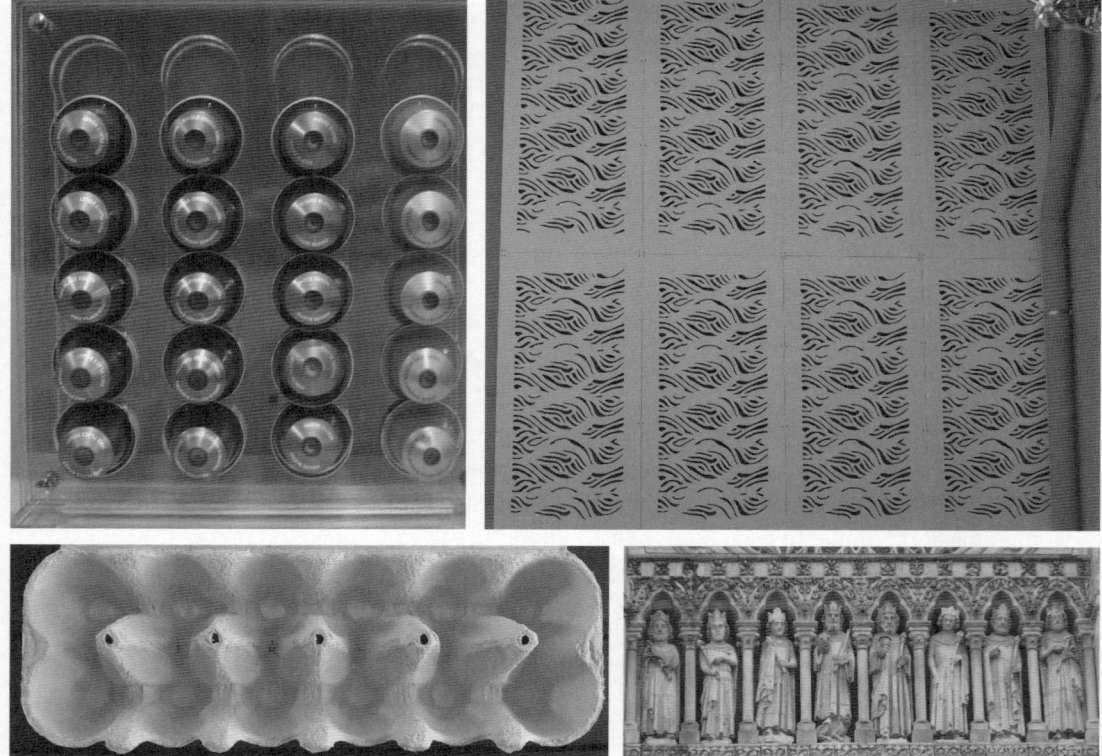

**Figure 4.4**   Examples of everyday arrays (5 × 4, 2 × 4, 2 × 6, 1 × 8)

In arrays, such as the coffee pod holder on the left of Figure 4.4, one can see how it could be used for skip counting in 4s or 5s and also for repeated addition of 4s or 5s, leading to ultimately seeing and knowing that 4 lots of 5 equals 20. By looking at the array from a different perspective, the commutative law of multiplication could be seen (4 lots of 5 is the same as 5 lots of 4).

There are many arrays available to be explored in the real world. Look at Figure 4.5. The first is an example of arrays in architecture – which can you identify? The second is a war cemetery – how does this represent an array?

**Figure 4.5**    Left: Château de Chenonceau; Right: War cemetery

If you are a teacher, you should find arrays great discussion starters with your class, but even better, they can be useful for practising personal number facts, especially for some of the trickier number facts (e.g. the 7 and 9 multiplication facts).

## Learning activity 4.8

In this learning activity we are asking you to take a photo safari! Look around your house (e.g. your kitchen or garden) or visit your park or local shopping centre or even walk around your local streets. Take photographs of different arrays and use them to practise or extend your knowledge of multiplication facts. Try to find some of the more uncommon arrays, such as arrays of 7s or 9s.

# PUTTING IT ALL TOGETHER: THE POWER OF NUMBER FACTS

Knowing number facts gives us great numeracy power. Having instant recall of number facts allows the focus of mathematical situations encountered to be on the processes required to solve them. It also gives the user improved confidence and competence to engage rather than avoid. Having a calculator with us (as in a mobile phone) can always help, especially with bigger or more complex calculations, but it is still worthwhile developing number fact automaticity. Sometimes, overusing our phone calculators for simple calculations can make us a little rusty on our number facts and get in the way of the development or retention of automaticity.

The knowledge of a number fact combined with an understanding of how our Hindu-Arabic number system works (such as knowing place value and what happens when we multiply or divide a number by 10) has amazing applications and lets us extend our ability to use number facts. Let's take the multiplication number fact of $7 \times 9 = 63$; here are some other things that are instantly known:

1. Commutative: $9 \times 7 = 63$
2. Related divisions: $63 \div 7 = 9$ and $63 \div 9 = 7$
3. Powers of 10: $7 \times 90 = 630$ (this works because $7 \times 90 = 7 \times 9 \times 10$); $70 \times 90 = 6\,300$; $630 \div 7 = 90$ (this can go on forever and you can see our Base 10, place value, and zero at work)
4. Decimals: $7 \times 0.9 = 6.3^*$; $63 \div 90 = 0.7$; $0.7 \times 0.9 = 0.63$ (this can go on forever as well and again, this is Base 10, place value, and zero at work)
5. Percentages (e.g. discounts): 70% of $900 = $630; $7\,000 \times 90\% = $6\,300$
6. Unit fractions: $\frac{1}{9} \times 63 = 7$; $\frac{1}{7} \times 63 = 9$

*Note: When multiplying by a number less than 1, as in Examples 4 and 6 the answer is less than the number we started with. For example, in $7 \times 0.9 = 6.3$, 6.3 is smaller than 7. This can sometimes be confusing as we initially learn that when we multiply, numbers get bigger and when we divide, they get smaller, but this is the opposite when dealing with decimals: for example, $63 \div 0.7 = 90$.

Although the list and examples could go on, the message is that knowing one number fact opens the gates to vast applications. In fact, we consider all the above examples as the same number fact; in each of them, the 7, 9 and 63 can be seen but they appear a little differently with zeros, or decimals or dollar signs or percentage signs. Imagine the vast possibilities of applications if all the number facts were considered and all were known with automaticity!

Knowledge of an addition fact can also have broader applications. Consider $6 + 8 = 14$; some possible uses include:

- commutative: $8 + 6 = 14$
- related subtractions: $14 - 8 = 6$; $14 - 6 = 8$
- larger numbers: $60 + 80 = 140$; $600 + 800 = 1\,400$
- decimals: $0.6 + 0.8 = 1.4$; $0.08 + 0.06 = 0.14$

## Learning activity 4.9

For each of the following calculations, use your number facts to find an answer without using a calculator or pen and paper. You can use your number fact grids if necessary. Remember to think about place value and fact families as you go.

At this point we want to look at some mathematical calculations involving procedures that allow us to apply our new skills with number facts. There are four operations (+, −, ×, ÷), each with its own unique procedures for calculating using pen and paper and in the following example we will only look at addition. We don't want to present mathematics as a series of procedures; however, it is important to understand that with practice and use of the number facts, such calculations are not as daunting as they might first seem. If you are a little rusty on some of these procedures, perhaps you could look at some of the many websites and online tutorials for some revision.

When faced with a calculation such as 145 + 89, few people are able to complete it easily in their heads, though there are strategies for doing so, which we will investigate in the next chapter where we look at mental computation. To complete this calculation with pen and paper, the numbers are first written in vertical format (with attention to place value) and then added, as shown in Figure 4.6.

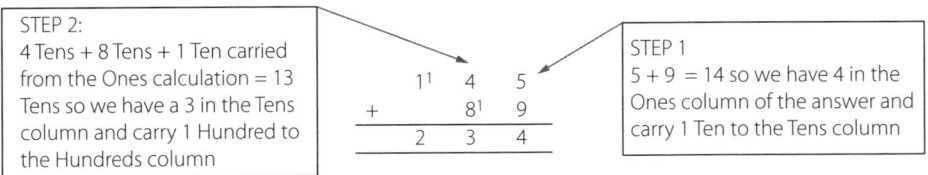

STEP 2:
4 Tens + 8 Tens + 1 Ten carried from the Ones calculation = 13 Tens so we have a 3 in the Tens column and carry 1 Hundred to the Hundreds column

STEP 1
5 + 9 = 14 so we have 4 in the Ones column of the answer and carry 1 Ten to the Tens column

**Figure 4.6**  Addition calculations of 145 + 89

Such procedures allow the Hindu-Arabic features to guide our calculations. Note that ultimately calculations such as that in Figure 4.6 are a progression of number facts. The addition of 145 + 89 used 9 + 5 and 8 + 4 in combination with the carried figures (i.e. 9 + 5 = 14) so the 4 goes in the ones column and the 10 is carried into the next column and becomes a 1 in the tens column. One can see in this example that knowing number facts with automaticity will allow a focus on the other processes involved, such as setting out and adhering to number system conventions. Conversely, not knowing the number facts would lead to considerably reduced speed, accuracy and confidence.

If you found the calculation in Learning activity 4.10 challenging, remember that many of us may not have had the need to do calculations using pen and paper for quite some time. Having a calculator on our phones has probably influenced this situation. We will revisit calculations such as these in future chapters so you will have more opportunities to practise and hone your skills again.

# RELATED SYSTEMS

We live in very fortunate times (mathematically speaking) as we are now very privileged to have other important systems in our lives that align beautifully with our number system and knowledge of number facts. In the relatively recent past, in Australia and other countries, the metric measurement system replaced the imperial system and decimal currency replaced the pounds, shillings and pence. These two new(ish) systems are built on Base 10, as opposed to the older systems; for example, 12 inches in a foot and 3 feet in a yard (imperial measurement) or 12 pence in a shilling, 20 shillings in a pound (old currency). Because our currency and metric measurement systems are Base 10 (and not Base 12, for instance), we can work with them using our knowledge of multiplication number facts up $10 \times 10$ and addition number facts up to $10 + 10$.

The history of the development of the metric system is a fascinating one as are the stories of the development of the old imperial system; it's definitely worth doing some online searches for these stories, especially if you are a teacher.

# CONCLUSION

In this chapter we focused on number facts. Knowing number facts with automaticity is vital for efficient and confident engagement in the numeracy elements of our lives. Number facts are at the core of our development of number sense and are a key element for the remaining chapters. So, just as al-Khwarizmi did many centuries ago, throughout this chapter and its learning activities you have taken knowledge of the number system, in particular knowledge of how the 10 digits relate to each other through addition, subtraction, multiplication and division (i.e. through the number facts), and developed the ability to instantly complete endless calculations.

## Personal actions to improve number sense

By following Polya's problem-solving strategy, it is possible to now *carry out the plan* to improve your personal number sense. The following are suggestions to assist you:

- Determine your automaticity of number facts: colour-code some grids. Be strict on yourself; automaticity means no thinking, just answer.

- Find and use online support, practice apps, singalongs, number facts quizzes and games.

- Deliberately pay attention to (collect with your phone camera) number fact situations, such as arrays or noticing signs indicating travel distances, interpreting timetables or infographics.

- Challenge yourself to work through mathematical situations based on number facts.

- Only use your phone calculator when you are really stuck or to check your answer.

- Give yourself some reminders to keep the number facts 'happening' as someone would in music practice, sport training or when learning a language.

- Take your time; any improvement is a triumph.

- Try not to let previous number fact experiences colour your future experiences.

- Remember unlearning something, especially an emotion, takes time and effort and can be frustrating.

# 5

# Mental computation

## LEARNING OBJECTIVES

After reading this chapter, you should be able to:

- use number facts and place value knowledge to develop mental strategies
- understand and apply properties of numbers that help with mental computation
- extend number facts and apply estimation to mental computation
- evaluate the reasonableness of answers and appreciate the need to do so.

# INTRODUCTION

In Chapter 3 we explored the Hindu-Arabic number system to understand its features and how they work and then in Chapter 4 we revised basic number facts and some properties of numbers and operations, which allow us to perform regular calculations. In this chapter we focus on how we use the knowledge of the previous two chapters in the most common way in our daily lives: mental computation. Remember that we have previously described mental computation as a key element of number sense. Much of this chapter will centre on the development of effective strategies for mental computation. We will look at the multiple ways in which many mental mathematics problems can be approached. In the latter part of the chapter we move from the application of number facts for mental computation to an important skill used every day by numerate people: estimation.

There are two reasons for a detailed focus on estimation: (1) it is important in everyday mathematical situations and problem solving and (2) it is important as a mathematical tool that allows us to judge the reasonableness of answers to mathematical problems (and sometimes the mathematical claims of others). In the next section we will look in detail at what is involved in mental computation and why it is so important in developing strong number sense.

# WHAT IS MENTAL COMPUTATION?

**Mental computation** puts into action our basic mathematical skills and understandings to help us with the majority of numeracy situations we encounter each day. Mental computation, as the name implies, is conducted 'in our head' without pen and paper or calculators. Computation hints at what

> **Mental computation** refers to calculations that are completed mentally – without the aid of physical or digital tools.

happens in our brain – it computes or acts like a computer. Mental computation is a key manifestation of our number sense; it is where we display our skills and understandings and where they come into play to make us functionally numerate. The better we are at mental computation, the quicker and more accurately we can make the multitude of daily decisions that can affect our lives in ways, from the insignificant (is there enough milk left for today?) to the life changing ('I know what you're thinking: "Did he fire six shots or only five?" These words were uttered by Clint Eastward in *Dirty Harry*. If you haven't heard of this movie scene, check it out on YouTube).

Mental computation can be required for many mathematical situations we encounter, so is very broad in scope, but for most of us to be functionally numerate most of the time there are some basic skills we need. These skills might include knowing our number facts, estimating (e.g. measurements, costs) and a range of mental calculation strategies. Mental computation is improved when we are able to:

- identify the appropriate operation(s) to use
- deal efficiently and fluently with numbers
- recognise patterns and relationships between and among numbers
- think strategically.

## Scenario 5.1

### EVERYDAY MENTAL COMPUTATION

We often use mental computation when we are in situations involving money. Here are some examples of such situations. Think about each and identify what you might calculate in each situation and the kind of calculation you might do.

**Figure 5.1**   Everyday situations in which mental computation might be useful

It should also be noted here that everyone from time to time encounters situations where our mental computation skills let us down (miscalculating exchange rates is our forte), so don't think that all situations encountered have to be 'done in your head', but it is handy to have the 'power' at hand whenever possible.

## Why the focus on mental computation?

Fascinatingly, for mental computation an accurate answer is not always required, although in the *Dirty Harry* scenario accuracy would be profoundly important. Often, for us to function smoothly in numeracy situations we just need an approximate answer: for example, 'Do I have enough fuel to get home?' requires a glance at the fuel gauge and a thought of the distance to travel. If the answer to this question is an obvious 'no' or 'too close to call', we will stop to refuel. Continuing this scenario, sometimes other numeracy complexities enter our mind, such as remembering that tomorrow morning you have to take your partner to the airport so may not have enough time to refuel on the way. In this scenario alone we would have engaged skills in time, distance and fractions or scale (fuel gauge). We all have had experiences when we have not thought these situations through and perhaps have been inconvenienced as a result. The fact remains that the capacity to use mental computation when appropriate is a life skill that is helpful and often time saving. It can sometimes even help us with critical orientation (see Chapter 2) because it lets us make judgements on the spur of the moment (e.g. Am I getting a good deal here? Will I have two or three scoops of ice cream?).

What is also intriguing about mental computation is that most of the time we are completely unaware that we are engaging in it. In the previous fuel gauge reading example, most drivers would not have even realised that they were looking at the gauge and were probably thinking fractionally: fuel tank $\frac{3}{4}$ full so continue to drive home, fuel tank $\frac{1}{8}$ full, maybe stop and refuel. To begin to analyse how we engage in mental computation, we have to deliberately become aware of it; we have to think about our thinking (**metacognition**). This is not always easy to do because, as stated, mental computation often happens without us consciously considering it.

> **Metacognition** means thinking about one's thinking.

---

## Reflection activity 5.1

Try now to review an hour of your life. Recall anything that you thought about with a mathematical basis and make a list. For instance, have you:

- checked the time?
- wondered how many kilojoules were in your snack?
- decided whether the groceries were too heavy to lift?

→

- calculated the cost of something?
- worried about your credit card interest?

Now that you have made a list, try to unravel the ways you thought about things: what mathematical skills did you engage? For the above situations it might look like this:

- Checked the time: Yep, I've got time to write another paragraph. (estimation of time and paragraph writing ability)
- Wondered how many kilojoules were in your snack: I won't have dessert this evening. (compensation strategy to balance kilojoules)
- Decided whether the groceries were too heavy to lift: OK, I'll make two trips to the kitchen. (halving the groceries and doubling the trips: inverse proportional thinking)
- Calculated the cost of something: Oh, I haven't got enough saved. I'll need to use my credit card. (subtraction: cost – savings)
- Worried about your credit card interest: Oh, I'll be paying 20% interest on this item for months/years. (percentages, time, money, opportunity cost)

By becoming a little more aware of your use of mental computation, you may begin to develop an insight into your own strengths and weaknesses, which again is the first step in Polya's problem-solving approach: understanding the problem. If you are like most people, including us, you probably identified times when you did try hard to address a situation with mental computation and perhaps there were times when you did not. Perhaps there were times when you/we have:

- avoided engaging in the situation (this happens a lot with credit card debt; people can just turn a blind eye or say *mañana*, overwhelmed by the urge to own their desired item)
- pretended something didn't happen (hid the snack wrapper, so it didn't really get eaten)
- deferred to someone else (how much do I owe you for lunch?)
- reached for a calculator (not always a bad thing but can become a bad habit).

With this metacognitive analysis (i.e. thinking about your thinking), hopefully your awareness has been raised, and we hope that by engaging with the ideas and activities in this chapter that two main things will happen to help you improve your mental computation; first, your use of non-engaging strategies (now identified) is lessened; and second, your day-to-day use of mental computation is highlighted so that you can consider and enhance your strategies for using this skill.

# USING NUMBER FACTS TO COMPLETE MENTAL CALCULATIONS

When we are young we are taught our number facts and then we begin to use the number facts to complete pen and paper calculations, such as 14 + 17 or 23 ÷ 5. When doing these pen and paper calculations, we are taught a procedure to follow and we mentioned this previously in Chapter 4 (see Figure 4.6). Here we will briefly revisit these procedures. Let's take the 14 + 17 example. In the vertical format shown below we add the 4 to the 7 first, get an answer, and then move on to the tens column. What is happening is that we of course are remembering our number facts (4 + 7 = 11) and then writing the answer down so that we don't have to keep the answer in our head. Note that in the calculation shown, the answer to 4 + 7 = 11 can be seen as a 1 in the ones column and the small superscript of 1 in the tens column.

$$
\begin{array}{r}
1 \quad 4 \\
+ \quad 1^1 \quad 7 \\
\hline
3 \quad 1
\end{array}
$$

When we have to do the same calculation in our heads, some interesting things happen. Some people report having a mental picture of the vertical format calculation and completing it with the power of knowing their number facts and remembering the steps of the calculation in their 'mind's eye': for example, remembering the answer for the ones column, the carry figure for the tens column, then the total for the tens column, and finally the overall answer. However, some people take different approaches.

---

## Learning activity 5.1

Before we continue, think about how you would mentally compute 14 + 17.

Interestingly, many people will abandon their pen and paper training of completing the ones column first and perhaps do something like the following examples. Which of these do you relate to?

- 10 + 10 = 20 then 4 + 7 = 11 then 20 + 11 (tens column completed first, then ones column, then combined)
- 10 + 10 = 20 then add 7 to get to 27 then add 4 (tens column completed first, then the largest ones column number added, then the smallest ones column number added)
- 14 + 10 = 24 then add 7 (maintain the first number, add the 10 followed by the 7 in the ones column)

The ideas in Learning activity 5.1 show that even for a simple calculation, several possible strategies can be used. Some people will use one strategy for most similar calculations; others will mix and match, flexibly, fluidly and confidently selecting strategies as the calculations vary. Which strategies you employ won't matter if your method is efficient and effective (i.e. quick and correct)! Developing efficient strategies can take a great deal of time and patience with yourself. Let's repeat the focus of Learning activity 5.1 with some multiplication calculations.

---

## Learning activity 5.2

1. Think about how you might go about calculating $3 \times 8 + 7 \times 8$
2. How about $25 \times 7$?

Do any of these ideas match with your thoughts?

1. $3 \times 8 + 7 \times 8$
   - You could calculate $3 \times 8 = 24$ and $7 \times 8 = 56$ and then add $24 + 56$ together to get 80 (but that may seem a little tedious).
   - Another approach is to say that 3 lots of 8 added to 7 lots of 8 is the same as 10 lots of 8 and therefore $10 \times 8 = 80$ (this uses the Distributive Law – see the next section).

   The second way is easier but requires experience and practice to recognise the strategy.

2. $25 \times 7$
   - $25 \times 7$ is the same as $7 \times 25$ (because of commutativity). I know that 7 lots of 25 is the same as 4 lots of 25 plus 3 lots of 25. My question now becomes $4 \times 25 + 3 \times 25$, which equals $100 + 75$, or 175 (this also uses the Distributive Law – see the next section).

---

# PROPERTIES OF NUMBERS AND OPERATIONS

In Chapter 4 we introduced commutativity and identities in relation to their usefulness when learning and remembering number facts. In this section we will look at some more properties of numbers and operations that can assist with mental computation.

## Order of operations

You may remember learning about the order of operations from school. This order is a convention agreed upon by mathematicians so that we all understand and use the same accepted order in which to make calculations. It is important to pay attention to the symbols and operations that you are using in any calculation (and also to remember that when you are using your calculator, it may not always attend to the order of operations). You may have used acronyms in school to help you remember the order (e.g. BEDMAS). The correct order is:

1. any calculation in brackets
2. any calculations with powers (e.g. $3^2$; powers are also called indices or exponents)
3. multiplication and division from left to right in the order in which they appear
4. addition and subtraction from left to right in the order in which they appear.

Here is a simple example:

$3 + 2 \times 4$ means we multiply first: $3 + 2 \times 4 = 3 + 8$, which equals 11. If we did not use the order convention, we would get a different answer: $3 + 2 \times 4 = 5 \times 4 = 20$. Mathematicians use conventions such as the order of operations to avoid the possibility of arriving at different answers. If the $3 + 2$ must be calculated first, then brackets are needed:

$(3 + 2) \times 4$ means we first calculate the brackets: $(3 + 2) \times 4 = 5 \times 4$, which equals 20.

## Distributivity

Did you notice that in Learning activity 5.2, the second strategy in Question 2 was similar to that in Question 1 but applied in reverse? These strategies rely on an important property known as the **Distributive Law**. In your days as a mathematics student, you may have seen it written using letters rather than numbers: for example, $a \times (b + c) = a \times b + a \times c$; or with subtraction in the brackets: $a \times (b - c) = a \times b - a \times c$. For many students, if this is their first experience with distributivity the abstract representation can make it difficult to understand. In our experience this symbolic representation also leads to students learning procedural approaches without ever understanding the concepts. A good way to remember the idea is to think of the everyday meaning of the word distribute (i.e. to spread or share). If you think about what's happening in the Distributive Law, the number outside the bracket is being 'distributed' across the terms in the brackets.

> The **Distributive Law** states that $a \times (b \pm c) = a \times b \pm a \times c$

Let's look at this idea at work using different representations, which are shown in Figure 5.2. The first two are written using numerical or symbolic notation. The third representation is a diagrammatic representation.

While we are not asking you to use the various representations shown in Figure 5.2, we hope that they help you to understand how and why the Distributive Law works. This knowledge and familiarity with its use can help to make your mental calculations more efficient. For example, the Distributive Law allows you to regroup the calculation so that you can use your number facts to help.

Please do not be discouraged if you didn't think of the ideas in Learning activity 5.2 or if these ideas still seem difficult to understand. We'll come back to them. (Remember your growth mindset!)

$$3 \times 2 + 4 \times 2 = \text{three lots of 2 added to four lots of 2}$$
$$= \text{seven lots of 2}$$
$$= 14$$

$$3 \times 2 + 4 \times 2 = (3 + 4) \times 2$$
$$= 14$$

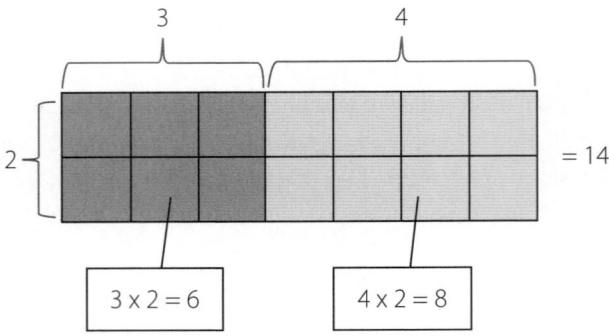

**Figure 5.2** Three different representations of the distributive law for $3 \times 2 + 4 \times 2$

## Learning activity 5.3

Apply the Distributive Law (and other properties such as commutativity if useful) to complete the following as mental calculations (you can check your answers with a calculator).

1. $4 \times 8 + 6 \times 8$
2. $3 \times 7 + 4 \times 3$
3. $8 \times 6 - 6 \times 6$
4. $\$25 \times 3 + \$25 \times 7$
5. $6 \times 15$ (hint: this strategy is the same as Example 2 in Learning activity 5.2)

*(Answers in Appendix.)*

In the next section we will look at another property that can be useful in mental computation.

## Associativity

You may remember our discussion in Chapter 4 about commutativity of addition and multiplication (useful because we know that $4 + 7 = 7 + 4$ and that $2 \times 3 = 3 \times 2$).

Another important property of addition and multiplication is **associativity**. In essence, this property tells us that when we are adding or multiplying more than two numbers, the order in which we carry out the operation doesn't matter. Let's look at some examples.

> The Associative Law (**associativity**) means that when adding or multiplying three or more numbers, the result will be the same regardless of how they are grouped.

## Example 1. Calculate without pen and paper or a calculator: $3 + 49 + 17$

Adding in the order as written, we calculate 3 + 49 (which equals 52) and then the calculation becomes 52 + 17. For some people this is not a challenge but for others it may be. We can use associativity to change the order and regroup the numbers to make the calculation easier: 3 + 49 + 17 = 3 + 17 + 49. Now calculating left to right first gives us 3 + 17 (which equals 20), and then 20 + 49 = 69.

## Example 2. Calculate without pen and paper or a calculator: $2 \times 13 \times 5$

Most people would find it tricky to do this calculation in the order that it is written because the first multiplication (2 × 13 = 26) means that the calculation would become 26 × 5. But if we know about associativity, we can regroup to 2 × 5 × 13, which becomes 10 × 13, and our knowledge of place value tells us the answer is 130.

The mathematical way of expressing the Associative Law is a + (b + c) = (a + b) + c or a × (b × c) = (a × b) × c. Again, remembering this abstract form of the property is not essential for number sense but when doing mental computation, knowing that we can 'associate' the numbers in a multiplication or addition to help us can be very convenient.

So, associativity is useful because it sometimes makes seemingly complicated mental computations easier by regrouping the numbers to let us apply convenient number facts and our knowledge of place value. *Remember*, though, this property *only* applies to addition or multiplication (not mixed together and never with division or subtraction – remember the order of operations). For example, it is not correct to say that (3 × 2) + 4 = 3 × (2 + 4) or that (12 ÷ 4) ÷ 2 = 12 ÷ (4 ÷ 2).

Up to this point we have presented you with a range of mental computation strategies. The next learning activity gives you the chance to select, practise, and reflect on your own strategies, after which we will share our ideas with you on what we have just covered.

# Learning activity 5.4

For each of the following calculations:

1. Complete each 'in your head' then record how you did it.
2. Now come up with a number of other ways you could have completed each.

3.  Now decide which is the most efficient and effective method for mental computation in each case.

4.  Make sure you read the explanations of these in the Appendix.

    a.  $26 + 34$
    b.  $47 + 38$
    c.  $65 - 24$
    d.  $71 - 28$
    e.  $23 \times 5$
    f.  $43 \times 8$
    g.  $147 \div 7$
    h.  $87 \div 3$
    i.  $97 + 112 + 3$
    j.  $4 \times 20 \times 15$

    *(Answers in Appendix.)*

If you had difficulty with any of the questions in Learning activity 5.4, it's important to analyse the problem: Was it with remembering number facts, holding the numbers in your head or deciding on a strategy? Remember, identifying a difficulty is the first step to addressing it. It may simply be that you have not thought about these calculations in this way before. Remember the discussion of growth mindset from Chapter 1? If you can't do these problems yet, don't be concerned: with time and patience, and a good deal of practice, you will be able to.

In recent years we have been doing some mental computation practice with our preservice teachers. A number of students always ask to write down the numbers (even though they are still calculating in their heads). It seems that they still want to be able to see the numbers while they do their mental calculations. This may be a supportive strategy that you could use. In the next section we will look at another important strategy that can be helpful for mental computation (as well as having many other applications): estimation.

## ESTIMATION AND MENTAL COMPUTATION

**Estimation** is the process of making a rough calculation of the amount or size of something.

**Estimation** can be a very useful tool in mental computation. According to the Oxford Dictionary, an estimation is 'a rough calculation of the value, number, quantity or extent of something' (Lexico, 2020). Because many of life's situations requiring mental computation don't need an exact answer, an approximate one will suffice. To achieve this, we employ the power of estimation, which in mental computation is all about making the calculation easier. For example, we may be happy to estimate the answer to $1.9 \times 8$ by knowing that it is approximately the same as $2 \times 8$ (which is 16).

Remember that reducing complex-looking calculations to something manageable is the key to success in mental computation. In the previous example we used a strategy known as **rounding** to change the value of 1.9 to an approximate value of 2. Rounding is one of the main strategies that can help us complete an approximate mental calculation and it is based on understanding the Hindu-Arabic number system. When we round we try to reduce the number of digits in a number to make it more usable. This requires us to consider place value. You may have heard instructions such as 'round to the nearest ten' or 'round to the nearest whole number'. Our rounding of 1.9 to 2 is an example of rounding to the nearest whole number.

> **Rounding** is making a number simpler while keeping its value close to the actual value.

Let's look at another example. It hurts to think about 409 ÷ 19 but a quick use of rounding to the hundreds or tens columns (place value) makes the calculation become 400 ÷ 20. At this point the calculation can be completed or further reduced in complexity by dividing both numbers by 10 to make the calculation 40 ÷ 2 (Base 10, place value). So, in a situation that doesn't require an exact answer, an estimate of 20 should be fine.

## Learning activity 5.5

Try your hand at using estimation in these calculations to find an approximate answer. Sometimes you may not find an easy way, which is okay – we can always revert to calculator or pen and paper – but attempting them mentally is important. As well as rounding, look for numbers that work well together (use your number facts). Try to figure out your own strategies before looking at the ones we offer. Remember the answers you obtain will be approximations of the actual answers.

a. 556 ÷ 68
b. 24 897 ÷ 508
c. 78 × 92
d. 41 × 0.29

Our strategies:

a. Round to 560 ÷ 70 then divide both numbers by 10: 56 ÷ 7 (number fact), so equals 8. By knowing the number facts, you can find them lurking in other numbers as in this example.
b. Round to 25 000 ÷ 500 then divide both numbers by 100: 250 ÷ 5 (number fact 25 ÷ 5 but 10 times bigger), so equals 50.
c. Round to 80 × 90; number fact 8 × 9 = 72 but both numbers are 10 times bigger so 100 times bigger overall: 72 × 100 = 7 200. Another way to think of it is that 80 × 90 = 8 × 10 × 9 × 10 so we know that this is the same as 72 × 100.
d. Round to 40 × 0.3. The number fact is 4 × 3 but the 40 is 10 times bigger and the 0.3 is 10 times smaller so they compensate one another, and the answer is 12.

In these examples the Hindu-Arabic number system can be seen at work throughout with the use of Base 10 understanding, place value, and zero. Again, this is why we emphasised the importance of knowing and understanding our number system as a key to developing number sense. Also realise that in each of these examples we were dealing with plain numbers. In life we are more likely to encounter these kinds of mental computations with money or measurements; the same strategies apply but sometimes need us to consider the reality of our answers. Let's look at an example in Scenario 5.2 where rounding needs to be carefully thought about.

## Scenario 5.2

### MAKING CURTAINS

When we use estimation, depending on the situation, it may still be important to be careful about how we round numbers. For example, if I want to estimate how much fabric I will need to make three curtains, each of which is 210 cm (2.1 m) long, rounding down to 200 cm (2 m) would give me an approximate answer of 600 cm (or 6 m). Can you see the problem? Six metres will only be enough to make three curtains that are each 200 cm long so my curtains will be too short.

### Learning activity 5.6

Reflecting on the situation in Scenario 5.2, what might be some other situations or contexts in which rounding while estimating might prove problematic?

## CHOOSING STRATEGIES TO AID MENTAL COMPUTATION

Once number facts are known to automaticity, and confidence and competence in using them have developed, the next step is to exploit these powers in diverse situations. So far, we have only considered mental computation in regard to completing what effectively are number fact calculations in our heads, but often these calculations are the easy part. As you may have felt in our discussion about the distributive law or associativity, quite often the tricky part of mental computation is deciding which strategy to use. Hopefully the discussion we have just had about rounding and the use of estimation will also give you some strategies to help.

We mentioned earlier that we are both prone to errors when calculating exchange rates; these errors don't come from being unable to multiply or divide in our heads, but from selecting the wrong strategy – for example, multiplying instead of dividing or miscalculating decimal situations. This can also depend on the exchange rate we are dealing with and how keen we are to know the exact value after conversion. Let's look at some examples in Scenario 5.3.

## Scenario 5.3

## CURRENCY EXCHANGE

One Australian Dollar buys 61 Euro cents. If an item costs 30 Euros, what will it cost (approximately) in Australian Dollars? Remember that an exact answer is not always necessary; we just want an answer close enough to know if the item being purchased is good value. To find the price in Australian Dollars, the calculation is 30 Euros ÷ 0.61, remembering that we are working with Euro cents, making the decimal necessary. Also remember that when dividing by a number less than 1, the answer will be greater than the initial number – in this case, greater than 30.

Possible solutions to get an approximate answer in our heads: We could round 0.61 to 0.6 and divide 30 by 0.6, but dividing by decimals can sometimes be challenging, so here are a couple of other options:

- Make both numbers 10 times bigger to get rid of the decimal (so we are dealing with whole numbers): 300 ÷ 6 = 50
- With some sleight of hand (or mind), the problem can be reduced to a number fact, 30 ÷ 6, and we then remember our place value to estimate that 30 Euros is approximately equivalent to 50 Australian Dollars.

Once this strategy is worked out at the beginning of a trip it can be replicated very easily for any other prices (assuming constant exchange rates): for example, 42 Euros (420 ÷ 6 so $70 approx.); 180 Euros (1 800 ÷ 6 so $300 approx.)

The example in Scenario 5.3 shows a number of strategies working together in our heads:

*Rounding:* Rounding the 0.61 Euro cents to 0.6 did wonderful things for mental computation; it changed a potential division by hundredths to one of tenths (much easier).

*Place value/Base 10:* Multiplying both numbers by 10 (from 30 ÷ 0.6 to 300 ÷ 6) maintained their relative values and moved the calculation into whole numbers (even easier).

*Number facts:* Recognising that the number fact of 30 ÷ 6 exists within 300 ÷ 6 (back to automaticity) and therefore guides the calculation.

Up to this point we have described estimation using rounding as a means to simplify calculations so that you can manage them in your head. Now we want to look at another important application of estimation and one that is a key skill used by someone with strong number sense.

# IS THE ANSWER TO MY MENTAL COMPUTATION REASONABLE?

One of the best determinants of a person's number sense is their ability to decide whether an answer is 'reasonable'. It is an extremely important element of any mathematical calculation (mental, pen and paper, or calculator) to check the answer to see if it is reasonable. As such, 'reasonableness' is one of our favourite mathematical terms.

As seen in the examples in the previous section, there often are numerous steps to take mentally to determine an approximate answer. An error in any of these steps will lead to an overall incorrect answer (again we're thinking of some of our disastrous exchange rates calculations over the years). So, checking the answer for 'reasonableness' is a vital final step. We often find that children using calculators have an overwhelming faith in the answer provided and do not think about its reasonableness. Errors in keying, decimals or choice of operation can lead to extraordinary answers. A young student of ours was using a calculator to work on a problem of the cost of petrol for a 200 km driving trip. The problem involved the distance, the cost of the petrol per litre, and the fuel consumption of the car in litres per 100 km. There were obviously numbers with zeros, costs with decimals (dollars and cents), and fuel consumption rates also involving decimals. The correct answer to the problem was $25, but the student's answer, because of incorrect input of decimal numbers, was $250 000. When asked to look at his calculator and think about the answer, he was adamant that it was correct (because that is what his calculator said). It was only when we asked, 'Do you really think it will cost a quarter of a million dollars to drive 200 km?' that the reasonableness (or unreasonableness) of his answer dawned on him.

The other unusual thing about 'reasonableness' is that often people with good number sense have an 'inkling' of the answer before the mental computation is started, which then simply makes the calculation a confirmatory process; the calculation reinforces the reasonableness. When completing a mental computation, always take a microsecond to consider the veracity of the answer. Is it possible? Does it make sense? Is it reasonable?

## Scenario 5.4

### CHECKING CHANGE

While using cash for purchases is growing less frequent, most of us still use it sometimes. When handing over an amount of cash it is a good habit to quickly estimate the amount of cash expected as change. For example, $20 cash minus cost of $12.20 (round to $12) so you'd expect about $8. We recently watched a bartender work a scam with change. We paid with a $50 note but received change for a $20 note. We realised this because we had estimated a certain amount of change and what we received didn't seem reasonable. We pointed out the problem and were given correct change. We noticed, however, that the same situation occurred with a number of other customers; the change given was for a lesser note. To exercise your number sense, make a point of mentally calculating an estimation of your change.

## Scenario 5.5

### COSTS AT THE CAFÉ

The situation represented in Figure 5.3 is a sign from a takeaway café.

**Figure 5.3**   A meal deal at a local café

We wondered what some possible prices for the rice and the three hot plate choices in Figure 5.3 might be. One possibility might be $7 for the rice and $1 for each plate. Is this reasonable?

Suggest some other possibilities for the prices in Scenario 5.5? Which answers seem more reasonable than others? Which seem unreasonable?

# CONCLUSION

In this chapter we have focused on mental computation – the process of calculating in our heads without the aid of such tools as calculators or pen and paper. While it is not always easy to do everyday calculations using mental computation, often it is more convenient. The efficiency and accuracy of mental calculations relies on the strategies that you choose to apply. In this chapter we have examined a range of strategies that we commonly use to assist us in mental computation. We hope that some of these strategies are useful for you and that, with time and persistence, you will develop confidence to use your mental computation skills. It is also often the case that an exact answer may not be necessary, and to assist with such calculations we focused on estimation and the use of rounding to help us reach an approximate answer.

## Personal actions to improve number sense

By following Polya's problem-solving strategy, it is possible to now *carry out the plan* to improve your personal number sense. The following are suggestions to assist you:

- Deliberately become aware of your use of mental computations and conversely your use of computation aids (e.g. calculators).

- Think about your thinking with mental computation. Consider the strategies covered in this chapter.

- Review your day. When did you need an accurate calculation and when did you need an approximation?

- Think of the times recently that you used the skill of estimation to make a mental computation easier.

- Identify times when rounding helped to make a mental computation easier. Were there times when rounding could have caused problems (e.g. the curtains were too short)?

- Have you used number properties (e.g. distributivity, commutativity, associativity) or order of operations to help with a mental computation?

- Try to use the terminology to label your thinking (e.g. estimating, rounding, associativity).

- Consider the reasonableness of your mental computations.

# Additive and multiplicative thinking

# 6

## LEARNING OBJECTIVES

After reading this chapter, you should be able to:

- understand the key ideas about additive and multiplicative thinking
- distinguish between situations of additive thinking and multiplicative thinking
- explain the difference between absolute and relative thinking
- understand how relative thinking and multiplicative thinking are connected.

# INTRODUCTION

The ability to decide when to use the various possible operations (+, −, ×, ÷) is another important element of number sense and is closely connected to our knowledge of number facts and the properties of operations (which were the focus of Chapters 4 and 5). Such decisions generally come down to a choice between two types of thinking: additive thinking (including subtraction) and multiplicative thinking (including division). These types of thinking are the focus of this chapter. While the choice of thinking might seem straightforward, making the shift from additive to multiplicative thinking is known to be a key reason for some students not making progress in mathematics, and this can persist well into the middle years of schooling (Dole et al., 2012; Hilton et al., 2012).

As we'll see in this chapter, there are times when distinguishing between additive strategies and multiplicative strategies can be tricky, so such decisions need some thought and an understanding of what distinguishes additive from multiplicative thinking. Once we have decided which thinking a situation requires, calculations can follow (either using mental computation or other approaches). It is important to persist with these ideas because understanding additive and multiplicative thinking (especially multiplicative thinking) is a cornerstone for the concepts involved in fractional thinking, and in understanding rate, ratio and scale (which are the subjects of Chapters 7 and 8). It is very difficult to isolate these concepts from each other so they may sometimes arise informally in this chapter.

The latter part of the chapter will build on the discussion of additive and multiplicative thinking to overview absolute and relative thinking, which are closely connected to additive and multiplicative thinking. They are also two of the more commonly used, but infrequently recognised, number sense skills. As with additive and multiplicative thinking, knowing when to use either absolute or relative thinking is an important ability.

# ADDITIVE AND MULTIPLICATIVE THINKING

Let's begin our discussion of additive and multiplicative thinking by developing an understanding of what these terms mean. We will start with additive thinking.

## Additive thinking

**Additive thinking** involves using counts when reasoning about quantities in ways that involve applications of counting, addition or subtraction.

**Additive thinking** involves using counts when reasoning about quantities – this usually means dealing with addition and subtraction (Bright, Joyner & Wallis, 2003; Hilton et al., 2016). Let's look at an example. Two teachers, John and Angela, are comparing the size of their classes. John has 28 children and

Angela has 25. Who has the larger class? While you weren't expected to calculate anything, you would most likely have used additive thinking to conclude that John has the larger class because you compared two quantities and considered the difference between them.

Different authors describe and define additive thinking (sometimes referred to as additive reasoning) in different and sometimes more detailed ways. For example, Siemon et al. (2011) described additive thinking as the connection between counting and part-part-whole and place value concepts. We will consider what it means to be thinking additively, especially as distinct from thinking multiplicatively as we move through the chapter. Let's continue our focus on additive thinking in Learning activity 6.1.

## Learning activity 6.1

In the following situations additive thinking is useful. Think about what operation you would use to find the answer to each.

1. Amy and Kate went shopping. They began with $20 and spent $12. How much money do they still have?
2. Kyle has been collecting toy cars. He has 34 of them. His grandmother has just arrived with his birthday present – another 6 cars. How many cars does Kyle have now?
3. Tran and Andy are having a dinner party. They have invited 6 guests. How many places will they need at their table?

*Question 1.* We know that Amy and Kate started with $20. If they spent $12, they must have $8 left (because 20 – 12 = 8).

*Question 2.* The number of cars Kyle has increased so we use addition: Kyle now has 40 cars because 34 + 6 = 40.

*Question 3.* Although we aren't given much numerical information, we know that there are 6 guests and we can reasonably assume that Tran and Andy will be joining their guests for dinner so altogether there will be 8 people (2 + 6 = 8) so they will need 8 places at their table.

For decades researchers have examined the ways in which children move from additive to multiplicative thinking. Sowder et al. (1998) found that additive thinking develops quite naturally in children. Perhaps this is because some of children's first experiences with numbers involve learning to count and comparing small quantities, which allows them to intuitively use additive thinking. Researchers have also found that for many people multiplicative thinking is not so intuitive and does not develop easily (Sowder et al., 1998; Hilton et al., 2016). It is important to be able to think multiplicatively because this is a key element of numeracy that has many applications in our everyday lives as adults. In the next section we will look at multiplicative thinking and how it differs from additive thinking.

## Multiplicative thinking

Although it is complex in nature and has application across a wide range of contexts, put

**Multiplicative thinking** involves applying concepts and approaches associated with multiplication and division to make comparisons between or among quantities.

simply **multiplicative thinking** involves using concepts and ideas related to multiplication and division (Siemon et al., 2011). Bright et al. (2003) defined multiplicative thinking as the comparison of quantities using ratios. It is reasonable to say that it is difficult to define multiplicative thinking in a succinct way because, as we will see, it is complex and varies depending on the relationships involved and the contexts or situations in which it is used. What we do know is that such thinking is essential for many mathematical processes, including:

- working with fractions, percentages and decimals (Chapter 7)
- dealing with situations involving ratio, rate, proportion and scale (Chapter 8)
- using relative thinking as distinct from absolute thinking (later in this chapter).

At this stage we would like to focus more on helping you understand what ideas are involved in multiplicative thinking than be overly concerned about definitions. In Learning activity 6.2, we look at some simple examples involving multiplicative thinking.

## Learning activity 6.2

In the following situations multiplicative thinking is useful. Think about how you would find the answer to each before reading our explanations.

1.  Bottled cola comes in packs of 6. If you want to buy 30 bottles of cola how many packs of 6 do you need?
2.  It costs $1.10 to post a letter. How much will it cost me to post 5 letters?
3.  For my holiday I have packed 4 different shirts and 3 different pairs of shorts. They are shown in this image.

How many different outfits will I be able to wear by pairing a shirt and shorts?

*Question 1.* This situation involves division. I know that I need to buy 30 bottles of cola so if each pack contains 6 bottles, I will need to buy 5 packs (because 30 ÷ 6 = 5).

*Question 2.* This is a situation involving multiplication. It will cost me $1.10 to post 1 letter so it will cost $5.50 to post 5 letters (because 1.10 × 5 = 5.50). You could also have used additive thinking by using repeated addition (1.10 + 1.10 + 1.10 + 1.10 + 1.10 = 5.50). Can you see how the additive strategy is more tedious, and if you were calculating the cost of posting 50 letters could become very time consuming?

*Question 3.* This is another multiplicative situation. For each pair of shorts I have a choice of 4 shirts. Since I have 3 pairs of shorts and can choose from any of the 4 shirts, I can make 3 × 4 outfits (i.e. 12 outfits).

Not all examples involving multiplicative thinking are simple and it may require some thought to decide on the appropriate approach. It is often easier to understand additive and multiplicative thinking by considering how we can distinguish between them. We look at this idea in the next section.

# DISTINGUISHING BETWEEN ADDITIVE THINKING AND MULTIPLICATIVE THINKING

As you will no doubt have gathered, selecting the correct strategy (either additive or multiplicative) is the start of a successful outcome. It is important to know when to use an additive strategy or a multiplicative one. Let's look at some examples.

## Scenario 6.1

### ADDITIVE OR MULTIPLICATIVE?

1. I want to make fruit drink for my friends. The label on the fruit juice concentrate container says that I will need 50 mL of fruit juice concentrate for every 250 mL of water. I plan to use 1 L of water and need to decide how much concentrate I need.

   This is a multiplicative situation because I need to keep the amounts of concentrate and water in proportion. (Imagine what would happen to the taste of my drink if I didn't.) Since 1 L is 4 times 250 mL, I need to multiply 50 mL by 4. I will need to mix 200 mL of fruit concentrate into my 1 L of water.

2. At the age of 12, Steve is twice his brother Andrew's age. Steve wants to know how old Andrew will be when Steve is 36.

While this situation may appear to be multiplicative, it is actually an additive situation because the gap between the brothers' ages will always be 6 years. So, Andrew will be 30 when Steve is 36. Some people may use multiplicative thinking to solve this problem and say that since Steve's age has been multiplied by 3, Andrew will be 6 × 3 = 18 years old. Can you explain why this incorrect thinking might happen?

When a situation is encountered that requires multiplicative thinking, it is often the case that things need to be kept in proportion, as was the situation in Example 1 of Scenario 6.1. Scenario 6.2 has another example of this type of situation.

## Scenario 6.2

## MIXING CONCRETE

Consider this scenario but please note we've made up these proportions – don't blame us if your fence post falls over!

I'm mixing some concrete, which requires water, cement and gravel. For every 4 bags of cement, I need 10 bags of gravel. If I make a batch of concrete with 6 bags of cement, I need to know how many bags of gravel I will need.

## Learning activity 6.3

Think about how you would answer the question in Scenario 6.2 and decide on your answer before looking at the following responses. We have described the responses we often get when we ask this question. When you read each response, try to think about your own thinking.

- *Answer: 12 bags of gravel.* These respondents saw that two bags of cement had been added (4 + 2 = 6) so they added two bags of gravel (10 + 2 = 12). The respondents knew that a change in bags of cement needed a change in bags of gravel. They used an additive strategy to arrive at their answer, but it is incorrect.
- *Answer: 15 bags of gravel.* These respondents saw that the required number of bags of cement was half as much again as the original number (i.e. it had been multiplied by one and a half times: $4 \times 1\frac{1}{2} = 6$), so they did the same for gravel: $10 \times 1\frac{1}{2} = 15$. Again, these respondents knew that a change in bags of cement needed a change in bags of gravel. They used a multiplicative strategy to arrive at their answer, which is correct. Note that the respondent may have approached the multiplicative situation slightly differently by increasing the amounts by 50%. The outcome is the same.
- *Answer: 10 bags of gravel.* These respondents did not think that a change in the number of bags of cement needed a change in the number of bags of gravel. They did not use additive or multiplicative thinking. The answer is incorrect.

In Scenario 6.2 the first step was recognising the situation as one where a change in one ingredient required a change in another. This thinking involves recognition of **co-variation**, which in this scenario is an understanding that an increase in one ingredient requires an increase in another. The third respondent did not recognise the co-variation needed between the amounts of cement and gravel. Respondents 1 and 2 recognised the co-variation between the amounts

> **Co-variation** is the situation in which as one quantity increases (or decreases) in value, the other quantity also increases (or decreases).

of cement and gravel but had to decide which strategy, additive or multiplicative, was needed. The key in this scenario is realising that the amounts of gravel and cement have to maintain their multiplicative relationship if the concrete is to be made correctly. This is the same thinking as in the first example of Scenario 6.1. In Scenario 6.2, in simplest terms, the 4 bags of cement and 10 bags of gravel can be expressed as follows: for every 2 bags of cement, 5 bags of gravel are needed.

The choice between additive thinking and multiplicative thinking in Scenario 6.2 may have seemed easy or obvious; however, a similar question to this was given to approximately 2 500 students in Years 5–9 (Hilton et al., 2013). Fewer than 25% of the students correctly chose a multiplicative strategy. Most students chose an additive strategy. The distribution of correct answers was similar across the grades, so no major improvement was apparent as the children got older. This finding is evidence that even well into the middle years of schooling, many students find multiplicative thinking challenging. Studies suggest that the majority of adults have similar difficulties (e.g. Lamon, 2007). In Scenario 6.3 we look at a situation in which being able to think multiplicatively is essential.

## A RECIPE FOR DISASTER

A tomato soup recipe for 6 people has to be reduced to cater for 3 people. Multiplicative Chef believes that to keep the soup tasting the same as the original, they must keep the ingredients in proportion by halving the original amounts. Additive Chef believes that if the number of people is reduced by 3, then each of the ingredients should also be reduced by 3. Their revised ingredients are shown in Table 6.1. Decide for yourself which chef's soup you would prefer to eat.

**Table 6.1**  Tomato soup recipes

| Original recipe for 6 people | Multiplicative Chef: 6 people ÷ 2 = 3 people | Additive Chef: 6 people – 3 people = 3 people |
|---|---|---|
| 10 tomatoes | 10 tomatoes ÷ 2 = 5 tomatoes | 10 tomatoes – 3 tomatoes = 7 tomatoes |
| 4 onions | 4 onions ÷ 2 = 2 onions | 4 onions – 3 onions = 1 onion |
| 3 cloves of garlic | 3 garlic ÷ 2 = 1 ½ garlic | 3 garlic – 3 garlic = 0 garlic |
| 6 teaspoons of salt | 6 tsp salt ÷ 2 = 3 tsp salt | 6 tsp salt – 3 tsp salt = 3 tsp salt |
| 3 cups of stock | 3 cups stock ÷ 2 = 1 ½ cups stock | 3 cups of stock – 3 cups of stock = no stock |

A couple of interesting things are illustrated in Table 6.1. The number of teaspoons of salt is correct in both methods. This can occasionally happen and can inadvertently reinforce the use of an incorrect strategy. For the other ingredients, however, things have become a little strange for Additive Chef to a point where the recipe would no longer taste much like the original. Notice that within Scenario 6.3 there were some simple addition and multiplication number facts. As emphasised in Chapter 4, knowing number facts is important but they then must be applied using correct thinking strategies.

## CAR BUDGETING

In this scenario we look at the application of additive and multiplicative thinking to the same situation. Jack is trying to decide how much he can afford to borrow to buy a car. He has found a car that he is keen to buy. The following list is his estimated car expenses for one year:

- Insurance: $1 530
- Maintenance: $750
- Fuel costs: $1 560
- Registration: $310
- Loan repayments: $5 800

Jack can calculate the total annual cost by using *additive* thinking (i.e. by adding the costs) to get a total of $9 950.

Jack finds it easier to budget if he knows what his average monthly cost will be. He can use *multiplicative* thinking to find out (i.e. by finding $9 950 ÷ 12). His average monthly costs would be $829.17.

On some occasions additive thinking is the simplest and most effective approach to use. Say I'm planting a punnet of 12 lettuces in my vegetable garden. I have dug 9 holes and planted 9 lettuces. To complete my planting task, I know I must dig a further 3 holes to cater for my 12 lettuces. This is a simple additive strategy and is the most effective to use. I could have complicated things a lot by using a multiplicative strategy:

- I have dug 9 holes and need to increase the number of holes by one-third ($\frac{1}{3}$ of 9 holes = 3 holes)
- I need to multiply 9 by $1\frac{1}{3}$ to get to 12

Sometimes we have to think additively and multiplicatively simultaneously when responding to a situation. A restaurant has a standard seating arrangement of 1 table and 2 chairs for couples. The restaurant is full but caters for a couple who arrive late. The owner brings out 2 more chairs (additive thinking: 2 more customers, 2 more chairs) and brings out 1 table (multiplicative thinking: 2 customers for each table).

## Learning activity 6.4

This activity involves using both additive and multiplicative thinking together. Table 6.2 contains costs for different levels of gym membership. Complete the final column to calculate the total cost of a six-month membership. When making your calculations, decide which part is additive and which is multiplicative.

*(Answer in Appendix.)*

→

**Table 6.2** Gym membership prices

| Program | Joining fee (paid once) | Monthly fees | Total cost for 6 months |
|---------|------------------------|--------------|------------------------|
| Bronze | $20 | $10 | |
| Silver | $30 | $20 | |
| Gold | $40 | $30 | |

Thus far we have looked at additive and multiplicative thinking. It is now important to address two other thinking strategies: absolute thinking and relative thinking. These are an integral part of our number sense and as we will see later in the chapter, they are related to additive and multiplicative thinking.

# ABSOLUTE AND RELATIVE THINKING

We examine absolute and relative thinking here because they use additive and multiplicative strategies. The meaning of the words *absolute* and *relative* in mathematical terms can be reasonably connected to their more general use in our lexicon. We use *absolute* or *absolutely* quite regularly. For example, if we say that someone has 'absolute power' we mean they have unchallenged authority; or if we respond to a request with 'yes absolutely', we mean 'without doubt'. Broadly speaking, both of these examples hint at a situation where nothing else is considered, and this is the context for

**Absolute thinking** is used when only one fact or quantity is considered.

**absolute thinking** in mathematics. The words *relative* or *relatively* have numerous usages in everyday language. For example, 'she is a relative' means someone in the extended family. 'The patient is doing relatively well' draws a comparison to their previous situation or to someone else in a similar situation. The first of these examples indicates a relationship and the second a term of comparison, both of

**Relative thinking** is used when two or more quantities are considered in a multiplicative relationship.

which point towards the mathematical meaning of **relative thinking**.

In mathematics, absolute thinking is used when only considering one fact. Relative thinking is when two or more facts are considered in a multiplicative relationship. However, like many aspects of our personal number sense, strict mathematical concepts are sometimes 'massaged' to suit the situation. A good way to remember these two methods of thinking is to consider the difference between price (absolute) and value (relative). When considering *price*, the actual cost is the only fact taken into account (and can often lead to disappointment – hence the expression 'buy cheap, buy twice'). When considering *value*, other facts may be thought of in conjunction with price (e.g. price and quality, price and aesthetics, or price and reliability). While these examples include some 'fuzzy' notions, which may vary from person to person (e.g. aesthetics can be subjective), in a strict mathematical situation the two facts or quantities that make up the relative situation are in a formal multiplicative relationship. Scenario 6.5 illustrates these situations.

## Scenario 6.5

## BUYING WINE

When buying wine we can use number sense to make decisions in a number of ways.

*Absolute thinking:* Buy the cheapest bottle; do not consider any other factors (many of us have been there!)

*Relative thinking without a strict mathematical formula:* Buy an affordable wine that you also consider to be good quality (thinking about two factors, price and quality). Remember that different people may have different opinions of quality and what they consider to be affordable, so this is not a strict mathematical relationship but still a valid thought process as part of our number sense.

*Relative thinking based on a multiplicative situation:* One bottle has a unit price of $20 but a case of 6 costs $96 so the unit price in a case is $96 ÷ 6 = $16. The decision to buy one bottle or a case can now be made remembering again that life is not always governed by strict mathematical thought – and you may not want six bottles, or you may not have $96.

## Scenario 6.6

## DESIGNER BRANDING

When a company wishes to market a product, it will often highlight points of differentiation from competitors' products. The idea is to entice the consumer to buy the product for some appealing reason (think cars, make-up or fashion). By developing a 'brand' the company is able to introduce other factors to customers who perhaps might only think absolutely about price, by encouraging them to think about price relative to the brand's 'special features'. The more enticing the brand is, the higher the price that can be charged because the non-price factors help to obscure the higher price. Designer handbags are a great example of this. High-end, well-known brands can charge thousands of dollars despite the absolute cost of manufacturing of the bag being only small relative to the final price tag. Can you think of examples of this 'branding' where your relative thinking was engaged, influenced or even manipulated?

Notice that when the comparisons were made in Learning activity 6.5 between the elephant and the car (relative thinking) it made it much easier in our mind's eye to understand the magnitude of the creature. Knowing the mass of the creature was 100 tonnes and 25 m long is amazing, but these are absolute amounts that we find difficult to imagine. Using the car and the elephant helps our understanding by giving us a point of comparison.

In our everyday lives there are situations that involve relative thinking but don't need mathematical calculations. We will now look at some scenarios that illustrate the recognition of relative situations or where we may or may not need to use relative thinking.

## Relative thinking and co-variation

Quite often in life an understanding of co-variation is all that is needed for our number sense to kick into action. Consider this example: It's going to be a very hot day so I will drink more water. There is no need to do a strict calculation here: we just need to know that the two factors, temperature and hydration, are involved and adjustments to our behaviour are required. In this situation an increase in one factor (temperature) led to an increase in another (maintaining hydration).

A decrease/decrease co-variation scenario could be from a doctor suggesting that a less sedentary lifestyle will lead to fewer health problems. Flipped around, this could be expressed as an increase/increase situation: increase exercise to improve health. Understanding co-variation is a very important part of our number sense and can have powerful effects on our day-to-day life. How often do we know the truth of a co-variation scenario but choose to ignore it (think about the doctor scenario above)?

## Learning activity 6.6

From your daily life, think of an example that requires understanding but not mathematical calculation for each of the following co-varying situations:

1. Increase/increase
2. Increase/decrease
3. Decrease/increase
4. Decrease/decrease

Now think about any co-variation scenarios that you are choosing to ignore (go on, be honest!). To start you off, here are two of our own: Increase/increase: Our weight always increases at Christmas because the amount we eat increases. Decrease/increase: Our bank balance always goes down around Christmas because of our increased spending.

*(Answers in Appendix.)*

## Co-variation in nature

Many creatures have an instinctive understanding of co-variation. Take this example: An Australian brush turkey builds a large mound of compostable material for a nest. The eggs in the mound are kept at a constant temperature range of 33 to 35 degrees Celsius. The male brush turkey regulates the temperature of the nest by adding or removing material. He is using co-variation to incubate the eggs: Add more compost to raise the temperature (increase/increase) or remove compost to lower the temperature (decrease/decrease).

Sometimes the quantities in a co-varying situation can change. A numbfish decided to eat a pufferfish with disastrous results, as shown in the photo in Figure 6.1.

**Figure 6.1**   A numbfish and a pufferfish dead on the beach

By instinct the numbfish knew that the size of its mouth relative to the size of the pufferfish would allow it to be swallowed. Unfortunately for both creatures, the puffer-fish changed its size by puffing up. Because the numbfish's mouth was not capable of co-varying in response to the pufferfish's increased size, the numbfish's 'relative thinking' no longer applied.

## Confusing relative thinking

Just as relative thinking didn't lead to a happy ending for the numbfish in the previous example, sometimes a situation that appears to be relative in nature can lead people to misjudge a co-variation connection. Scenario 6.7 and Learning activity 6.7 illustrate examples of this. In each case think about the situations in terms of co-variation and relative thinking.

### Scenario 6.7

#### MAGPIES AND CYCLISTS

Magpies become very territorial and protective in the springtime and swoop on the heads of passers-by. We recently saw a cyclist protecting himself from a magpie 'attack' by having inserted drinking straws in his helmet. The straws are meant to keep the birds at least some distance from one's head. This particular cyclist had inserted a large number of straws into his helmet – many more than the usual number, which appears to be about six.

**Figure 6.2**   Cyclists often use drinking straws to protect themselves from magpie attacks

**Figure 6.2** *(cont.)*

Consider these questions:

- Does the number of straws used represent a situation of co-variation with protection from magpies?
- Is there an ideal minimum number of straws that is sufficient for protection?
- Will adding more straws to the helmet add greater protection?
- Is this a relative situation; will doubling or trebling the number of straws double or treble the protection?

## Learning activity 6.7

A couple were staying with friends. One night, one of the guests woke with a very bad headache. His partner found some headache medication (for which the usual dosage was two tablets) in the bathroom cabinet. Without reading the dosage on the package, she gave the man two tablets as she usually did when he had a headache. They later discovered that the tablets had been double strength.

1. How many tablets should the woman have given her partner?
2. Some people might be tempted to think that taking twice as much medication would be twice as effective. Why should this situation not be thought of as relative?

*(Answers in Appendix.)*

The challenge for us all is to recognise when situations require absolute thinking, when they require relative thinking, when calculations are required and when they are not. Hopefully in Learning activity 6.7 you identified that the situation was not relative and that it would be dangerous to vary the dosage of any medication in the hope of changing its effectiveness. In the next section we will ask you to apply your understanding of absolute and relative thinking to look at some situations involving absolute and relative thinking with calculations.

# MAKING CONNECTIONS

Once a person is familiar with each type of thinking (multiplicative, relative, additive, absolute), the important thing remains to have the number sense to know which is appropriate to apply. Remember our previous discussion about co-variation? An ability to recognise situations of co-variation allows us to determine when relative or multiplicative thinking may be required.

## Connecting relative thinking to multiplicative thinking

In the previous examples we looked at the distinction between situations that are absolute and relative. We have seen that, sometimes, understanding a relative situation does not require formal calculations but when the situation is identified as relative, mathematical calculations can help us to make sense of the situation, make a judgement or draw conclusions. When this occurs, multiplicative thinking is required. In Learning activity 6.8 we look at some examples of this.

## Learning activity 6.8

1.  Look at the photograph of the flathead.

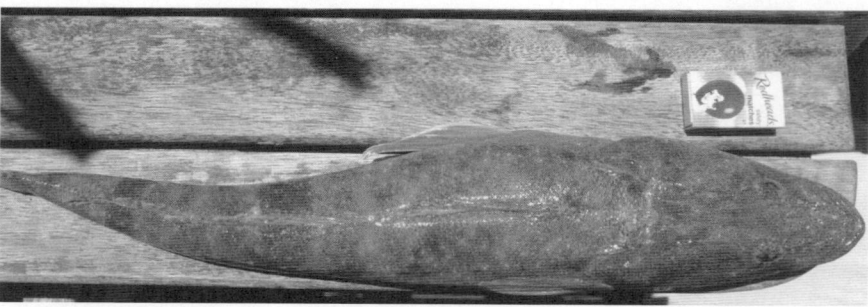

a.  Why do you think the fisher placed a matchbox in the photo?
b.  What type of thinking did the fisher invoke?
c.  Knowing that the length of the matchbox is approximately 6 cm, calculate the approximate length of the flathead.

**2.** Look at the two photos of statues:

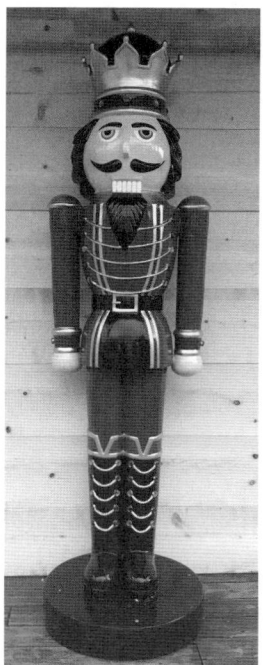 

**a.** What do you estimate the height of each to be?

Now look at the same statues in the following photos:

→

The first pair of photos shows only the statues, while the second pair shows the statues with a person next to them. Just looking at the first pair of photos, it would be difficult to determine the heights of the statues because there is nothing to relate them to. In the second pair of photos, relative thinking can be used because a person is present.

**b.** Knowing the height of the person is 2 m, what do you calculate the height of the statues to be?

*(Answers in Appendix.)*

In the examples in Learning activity 6.8 you used multiplicative thinking in a relative situation. In the following section we will draw on the ideas explored throughout this chapter to help you consider the concepts related to additive, multiplicative, absolute and relative thinking.

## Connecting additive to absolute and multiplicative to relative thinking

In this section we will use the context of the sinking of the *Titanic* to illustrate the ideas in this chapter.

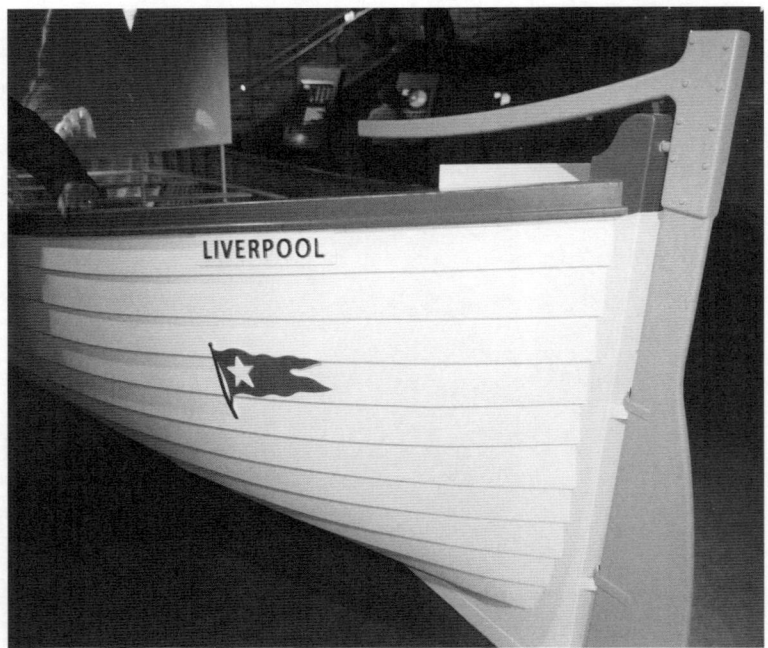

**Figure 6.3**   A *Titanic* lifeboat

Some of the lifeboats on the *Titanic* were not filled to capacity. This in part contributed to the high number of deaths. In Table 6.3 we show the numbers of children, women and men that were aboard the *Titanic* and we allow you the space to calculate how many in each group died.

**Table 6.3** Total children, women and men aboard the *Titanic* and numbers who died

| People | Children | Women | Men |
|---|---|---|---|
| On board | 109 | 425 | 1 690 |
| Saved | 56 | 316 | 338 |
| Perished (absolute number) | 109 − 56 = _____ | | |
| Percentage perished (relative to those on board) | _____ ÷ 109 × 100 = _____% | | |

# Learning activity 6.9

Look at Table 6.3 and answer these questions.

1. How many people were on board in total? (additive thinking)
2. Which group (children, women or men) had the highest number of people saved? (absolute thinking)
3. Complete the 'Perished' row of the table to show how many of each group perished. (additive thinking – using subtraction)
4. Complete the 'Percentage perished' row (we've started the calculations for the children data to help you – relative thinking).
5. Using your answers in the 'Perished' row, complete the following sentences:
   a. In absolute terms there were more _____ saved than the other groups.
   b. In relative terms there were more _____ saved than the other groups.
   c. In absolute terms more _____ perished than the other groups.
   d. In relative terms more _____ perished than the other groups.

Questions 5(c) and 5(d) had the same answer whether we thought about them absolutely or relatively. From your calculations, do you believe this was indeed a case of 'women and children first'?

*(Answers in Appendix.)*

Table 6.4 shows the number of passengers holding second- and third-class tickets on the *Titanic* as well as the number who were saved.

**Table 6.4** Numbers of second- and third-class ticket holders aboard and saved

| Passenger class | Passengers on board | Passengers saved | Passengers lost |
|---|---|---|---|
| Second | 285 | 118 | |
| Third | 706 | 178 | |

1. Using additive thinking, complete the final column of Table 6.4.
2. Using multiplicative thinking, calculate the percentage of second- and third-class passengers saved to allow us to make a relative comparison. (Do this by dividing passengers saved by passengers of that class on board and multiplying by 100.)
3. Complete these sentences:
   a. In absolute terms there were more _____ class passengers saved.
   b. In relative terms there were more _____ class passengers saved.
   c. In absolute terms more _____ class passengers perished.
   d. In relative terms more _____ class passengers perished.

   *(Answers in Appendix.)*

Although this section was quite short, we will revisit these ideas in Chapters 7 and 8. The important learning from the section is that absolute thinking considers only one quantity, whereas relative thinking requires us to consider two quantities.

# CONCLUSION

This chapter has looked at additive thinking and multiplicative thinking, both of which are important for reasoning and problem solving. These two forms of thinking allow us to use our number facts and to choose and apply operations to answer questions. We also looked at co-variation and described situations that require general number sense and others that require formal mathematical calculations. The chapter also looked at absolute and relative thinking with a similar focus on whether or not formal mathematical calculations apply. These important elements of mathematical thinking are strongly influenced by the contexts in which we use them, and it is not always straightforward to identify which is useful. In the final section of the chapter we drew on all of these ideas to think about a context (the *Titanic*) in different ways. We will continue to build on the ideas in this chapter because the concepts we have looked at here are also integral to understanding fractions, ratio, rate and scale, the focus of the next two chapters.

## Personal actions to improve number sense

By following Polya's problem-solving strategy, it is possible to now *carry out the plan* to improve your personal number sense. The following are suggestions to assist you:

• Be conscious of when you are using additive and multiplicative thinking.

• Identify situations that involve co-variation but don't require calculation.

• Identify co-varying situations that do require calculations and decide whether they are additive or multiplicative.

• Identify absolute and relative situations.

• If you identify relative situations, think about where multiplicative thinking would be useful.

# 7 Fractional thinking

## LEARNING OBJECTIVES

After reading this chapter, you should be able to:

- understand the structure of common fractions, decimal fractions and percentages
- perform simple calculations using common fractions, decimals and percentages
- describe the different interpretations of common fractions
- apply your understanding of common fractions, decimals and percentages to everyday contexts.

# INTRODUCTION

In this chapter we will look at fractions and what it means to engage in fractional thinking. Before proceeding further, please be assured that we know that the word 'fractions' is enough to cause concern for some people – and we have worked with many people who have overcome their initial trepidation towards fractions. The reality is that we encounter fractions a great deal in our daily lives and, as such, understanding fractions is another key element of number sense. Working to strengthen our understanding of fractions provides an opportunity to build our personal number sense.

We usually first encounter fractions in primary school, where we learn about fractions through numerous applications, some of which are rarely put to use in real life. So if your concerns about fractional thinking are dominated by experiences of calculations such as $2\frac{3}{4} \times 5\frac{1}{3}$ or $\frac{3}{5} + 2\frac{5}{9} + \frac{5}{12}$ or $\frac{7}{9} \div \frac{5}{6}$ or $\frac{3}{7} - \frac{2}{9}$, then relax because this is not what will be covered in this chapter. The aim here is to address some core concepts that will help foster functional number sense with fractions without getting too complicated. Thinking fractionally can be harder to visualise than whole-number thinking so it might be handy to grab some simple hands-on materials like buttons or counters to help work your way through certain scenarios or learning activities.

Fractional thinking can involve common fractions, decimals or percentages. As mentioned already, many people report difficulties with fractional thinking, and some identify it as the core reason for their lack of mathematical confidence. In this chapter some of the commonly identified difficulties are discussed, which we hope will allow you to relate to your personal experiences. Fractional thinking can doubtless become complicated, but in our day-to-day lives most of the numeracy situations involving fractions can be met with an understanding of a few basic fraction concepts. In school we probably encountered rather complex fractional algorithms, which are important for deep understanding of fractions but not so necessary for personal functional competence. To ground the chapter in personal number sense, images and scenarios from everyday situations will provide opportunities to illustrate and apply the fractional concepts.

This chapter is focused on three types of fractions. The three main ways to think fractionally is with common fractions (e.g. $\frac{3}{4}$), decimal fractions (e.g. 0.75) and percentages (e.g. 75%). We will start with common fractions.

# COMMON FRACTIONS

The definition of a fraction that most people are aware of is that it is a 'part of a whole'. This is a reasonable working definition, so we will keep working with it. This section deals with **common fractions** (called 'vulgar fractions'

> **Common fractions** are numbers that are expressed using two whole numbers in the form $\frac{a}{b}$, as shown in Figure 7.1.
>
> For example: $\frac{1}{2}, \frac{3}{4}, \frac{11}{20}, \frac{12}{7}$

in the past). A fascinating experience we have with many students is that they are able to identify the parts of a common fraction, but don't know what the words mean. When asked what the parts of a common fraction are, most students can recall the words 'numerator' and 'denominator', and some remember 'vinculum'. These elements of a common fraction are shown in Figure 7.1. Just like our experiences with learning about the Hindu-Arabic number system, it is important to understand what the words in Figure 7.1 mean as this tells us the function of each of the elements.

**Figure 7.1**  The elements of a common fraction

Consider the common fraction $\frac{3}{8}$. The 3 is the numerator, the 8 is the denominator, and the line, which should be horizontal but is made oblique (/) by most computers, is the vinculum. Both numerator and denominator are words derived from Latin. They end with an 'or' suffix, which means a person or thing that performs an action; for example, a sailor is *one who sails*. Therefore, an easy way to remember the meaning and function of each number in a common fraction is as follows: a numerator is *one that numbers* (you can see the link between numerator, numeral, number) and a denominator is *one that names* (you can see the word 'nominate' inside denominator, or the French word 'nom', meaning 'name'). We can think about the meaning of $\frac{3}{8}$ this way: the name of what we're working with is 'eighths' and the number of eighths we want is three. The word 'vinculum' also has its origin in Latin, meaning 'bond'. The use of the vinculum in mathematics can be complex and extend beyond simple common fractions. In the contexts in which we are working, the vinculum functions to separate the numerator and denominator.

We use the concept of number and name all the time in non-fractional situations: for example, I caught 3 fish (3 is the number and fish is the name); I ate 5 chocolates (5 is the number and chocolate is the name). Imagine a pizza cut into eight equal pieces (each piece is named an 'eighth'). Now imagine you want $\frac{3}{8}$ of the pizza, which means you want three of the eight pieces of the pizza. We could represent this as shown in Figure 7.2.

An interesting quirk with common fractions is that the order in which the numerator and denominator are said is not the order in which they are used. Take, for example, the $\frac{3}{8}$ of a pizza from the previous paragraph. If someone asked you for $\frac{3}{8}$ of a pizza, the first thing to be done is cut the pizza into eight equal pieces and then give the person three. So, while we say the numerator first, in this situation we use the denominator first. This quirk alone may have caused you as a young learner some difficulties as you were used to working with things in order.

**Figure 7.2**   A pizza cut into eighths, showing $\frac{3}{8}$

While on the topic of pizzas, another simple contradiction arises for children, in that they are taught that fractions must be in equal parts, but many of the real-life examples we use with them are not 'exact' – for instance, pieces of pie, pizza or cake. However, these are the sorts of things that we encounter in our everyday lives, so some flexible thinking is needed.

Before we continue with our focus on the more common uses of fractions in our everyday lives, we want to look briefly at some of the ways in which fractions can be used. It may be that some of these diverse uses of fractions have caused you confusion in the past. Our goal here is to provide a complete picture but we also want to keep things brief and as simple as possible. We ask that you continue to use your growth mindset and reflect on which of these aspects you feel confident about.

## Common fractions are versatile

A fraction can be thought of as part of a whole, but this is only one of several ways that we can interpret a fraction. These interpretations and uses of common fractions are described briefly in this section.

### Part of a whole

Common fractions can represent a part of a whole – for example, part of a cake or a rectangle. This is the fraction representing the portion of one unit (i.e. the whole shape or object). This representation is shown in Figure 7.3. Note that the denominator denotes the number of pieces in the whole (fourths or quarters) and the numerator tells us how many pieces are shaded.

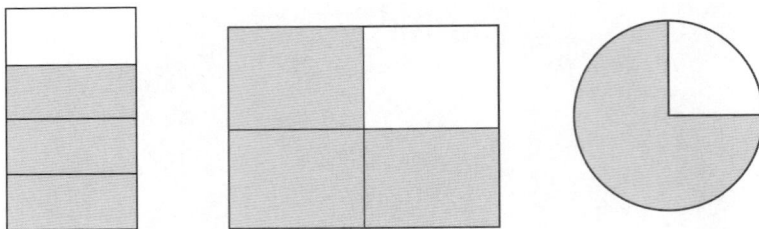

**Figure 7.3** The shaded sections represent $\frac{3}{4}$ (three-quarters of each whole shape)

It is important to consider what the 'whole' is, be it a number, a shape, a collection. In each of the examples in Figure 7.3 the whole is different, but we have shaded three-quarters of each.

## Learning activity 7.1

Complete the following instructions to practise using common fractions as part of a whole.

1.  Draw a rectangle and shade it to show $\frac{5}{6}$

2.  Draw a circle and shade it to show $\frac{1}{4}$

    *(Answers in Appendix.)*

Note that for each of the questions in Learning activity 7.1, you would have divided the shape equally to suit the denominator and shaded to show the numerator.

## Part of a collection

Common fractions are also useful for representing a portion of a collection – for example, a plate of sweets, a bag of marbles, a group of counters. In these situations, the collection is divided up into equal amounts (or sub-collections or sub-groups as indicated by the denominator) and the appropriate number of these sub-groups (as indicated by the numerator) is selected. This is shown in Figure 7.4 for the fraction $\frac{2}{5}$ (which can also be expressed as $\frac{4}{10}$). In this example, the collection of 10 counters was divided into 5 equal groups (each containing 2 counters). We then selected 2 of those groups to represent $\frac{2}{5}$.

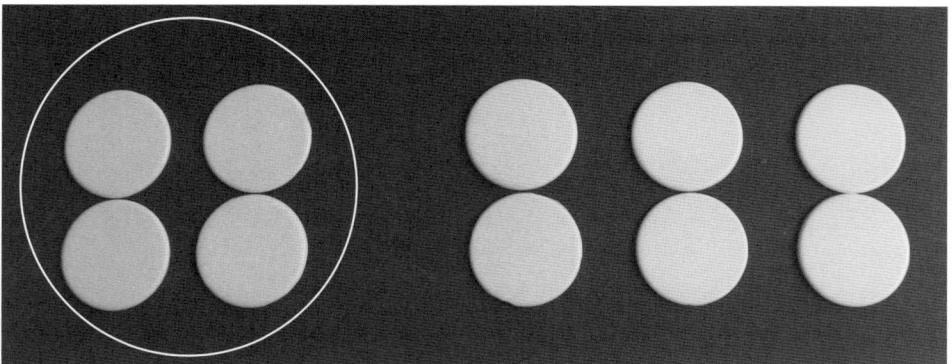

**Figure 7.4** Two-fifths of a collection

## Learning activity 7.2

Using some counters or buttons (or improvise with pieces of pasta), make the following collections, and then show the designated fraction (as we did in Figure 7.4).

1. Make a collection of 10 and then show $\frac{3}{5}$

2. Make a collection of six and show $\frac{2}{3}$

   *(Answers in Appendix.)*

Note that in each of the questions in Learning activity 7.2 you would have broken the collection into the number of (equal) groups indicated by the denominator and then selected the number of groups indicated by the numerator.

## A fraction as a point on a number line

A common fraction in which the numerator is smaller than the denominator is called a **proper fraction** and it is a number between 0 and 1. Because fractions are numbers, they can be placed on a number line. Figure 7.5 shows the position of $\frac{2}{5}$ on the number line. Note that the unit from 0 to 1 has been broken into five equal segments and where $\frac{2}{5}$ is located.

> A **proper fraction** is a common fraction in which the numerator is smaller than the denominator: for example, $\frac{2}{5}$. Its value is between 0 and 1.

**Figure 7.5** A common fraction can be represented as a point on a number line

A fraction where the numerator is greater than the denominator – for example, $\frac{4}{3}$ – is greater than 1 and is called an **improper fraction**. For example, if I ate $\frac{4}{3}$ of a pizza, it means I ate 1 whole pizza and an additional $\frac{1}{3}$ of a pizza, so altogether I ate more than 1 pizza.

## Learning activity 7.3

Draw a number line showing 0 to 1 (as in Figure 7.5). Locate the following fractions:

1. $\frac{1}{2}$

2. $\frac{2}{3}$

*(Answer in Appendix.)*

### A fraction as a division

Before we start to consider what it means to think about a fraction as a division, let's think about an everyday scenario.

## Scenario 7.1

## SHARING CHOCOLATES

Four friends want to share 3 chocolate bars equally among them. How many pieces should each chocolate bar be divided into? (Hopefully you said 4.) If the 4 friends now share the pieces equally among themselves, what fraction of a whole chocolate bar will they each get? (Hopefully you can see that each will get $\frac{3}{4}$ of a whole chocolate bar.)

We can represent this visually, as shown in Figure 7.6.

**Figure 7.6**   Sharing 3 chocolate bars among 4 friends

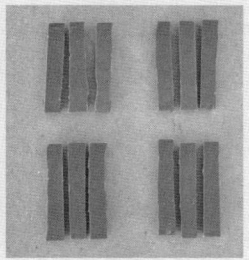

**Figure 7.6** *(cont.)*

Note that in Figure 7.6 the second photo shows each of the 3 chocolates broken into 4 equal pieces. The third photo shows the pieces rearranged so that each of the 4 friends will receive the same share (i.e. 3 of the pieces or $\frac{3}{4}$ of one of the original chocolate bars).

Scenario 7.1 is an example of a fraction working as a division. In this case we have 3 chocolate bars divided among 4 people (i.e. $3 \div 4$ or $\frac{3}{4}$). Notice that the vinculum acts in the same way as the division sign. Knowing this is very useful when it comes to doing less straightforward whole-number divisions in our heads because you can change the division sign to a vinculum. For example, $3 \div 7$ is the same as $\frac{3}{7}$. Understanding this makes problems like that in Scenario 7.1 easier. For example, dividing 5 chocolate bars among 7 friends would mean that each friend would receive $5 \div 7$ or $\frac{5}{7}$ of a chocolate bar. In Learning activity 7.4 we practise representing divisions as fractions.

# Learning activity 7.4

Thinking of the vinculum as a division sign, complete the following:

1. $5 \div 9 =$
2. $23 \div 37 =$

*(Answers in Appendix.)*

## Fractions to represent ratios

The easiest way to think about this situation is to consider another real-life scenario.

## Scenario 7.2

### COMPARING POCKET MONEY

Charlie receives $4 per hour from her parents for babysitting her brother. Patrick also receives money from his parents for helping with the housework. He receives $5 per hour. One day at school they compare their pocket-money payments. Charlie says it's not fair because she only gets $\frac{4}{5}$ of the amount that Patrick receives.

Let's think about the claim made by Charlie in Scenario 7.2. Is she right? We can think about the scenario by comparing the ratio of Charlie's pocket money to Patrick's pocket money – that is, $4 compared to $5. Another way to say this is that Charlie gets four-fifths of what Patrick gets. This means that Charlie's claim is correct.

### Learning activity 7.5

Thinking about Scenario 7.2 again, Charlie's parents have increased her pocket money to $6 per hour. Patrick's parents have not. If Patrick uses the same logic as Charlie did, what would Patrick say about this?

*(Answer in Appendix.)*

In this section we wanted to remind you of the ways in which fractions can be used. Having said that, most people only apply some of these in their everyday lives. Remember that the goal here is to improve your number sense, not make you a fractions expert! In the next section we look at the usefulness of common fractions.

## Why are common fractions useful?

In addition to the many ways that a fraction can represent mathematical relationships, the power of common fractions comes from their ability to be incredibly precise. If we think about it, there are an infinite number of possible numerators and denominators, which allows for many amazing fractions (e.g. $\frac{358}{2\,745}$). While this is a powerful attribute, it is far too complex for us to envisage, and not of much use in our day-to-day lives. For fractions to be a part of our everyday number sense, we need to focus on the most regularly used common fractions.

A number of common fractions have become part of our language. For example, most of us would be quite familiar with the use of fractions such as $\frac{1}{2}$, $\frac{3}{4}$ or $\frac{3}{8}$ when we chat about everyday quantities, such as half a glass of milk, three-quarters of a tank of petrol or three-eighths of a pizza. It is from these types of fractions and not the complex algorithms of our school days that we can develop our number sense. It is this kind of common fraction that will be used to illustrate some of the important fractional concepts in the rest of our focus on common fractions. We will now look at some operations with fractions.

## Operations with common fractions

In this section we look at some operations that involve common fractions – and we hope that in the process some of the challenges of the past may be lessened.

### Adding fractions

Now we know that the 'denominator' is the 'one that names', it helps to understand an important idea, which is: it is easy to add things together if they have the same name but difficult if they do not. Adding 2 bananas to 3 bananas = 5 bananas: the answer requires you to add the numbers and keep the name (we don't add the names). Now, if we try 2 bananas + 3 apples, the different names pose a problem when trying to express the answer. With common fractions, adding two fractions with the same name (denominator) is simple but imagine if we want to add two-thirds of a pizza to one-sixth of a pizza. The names (thirds and sixths) are not the same. If fractions have different names, then further thought is needed.

We did promise that this chapter would not focus in depth on calculations with common fractions, but we want to look at the following examples involving addition of common fractions because they provide important insights. (They are also more commonly encountered than subtractions, which work on the same principles.) Knowing about adding fractions can sometimes be useful, as shown in Scenario 7.3.

## Scenario 7.3

## FRACTIONS IN THE KITCHEN

Sometimes a recipe calls for $\frac{3}{4}$ of a cup (e.g. of sugar). I don't have a $\frac{3}{4}$ cup measure. I have a $\frac{1}{4}$ cup and a $\frac{1}{2}$ cup. I can either use my $\frac{1}{4}$ cup three times to get $\frac{3}{4}$, or I can use my knowledge of adding fractions. I know that $\frac{1}{2}$ is the same as $\frac{2}{4}$. I also know that $\frac{1}{4} + \frac{2}{4} = \frac{3}{4}$. This knowledge allows me to use my $\frac{1}{4}$ cup plus my $\frac{1}{2}$ cup to measure out $\frac{3}{4}$ of a cup of sugar.

In Scenario 7.3, when using the quarter cup and half cup, we had to change the name of the half cup to $\frac{2}{4}$, which allowed the numerators to be added. Note that a major stumbling block for children is that the names (denominators) are not added, only the numbers (numerators). This is tricky for young children who bring their embedded whole-number thinking to fractions. This is why it is so important to give them lots of hands-on experience where they can make and see the fractions in action. Advancing too quickly to a written fractions addition algorithm can lead to long-term confusion (which is what many adults report to us as having happened to them). This also highlights the importance of understanding the meanings of the terms *numerator* and *denominator*; as mentioned, many people know these terms but do not know their function and therefore do not understand their fractional application.

A disruption to understanding addition of fractions is sometimes encountered by children in school. For example, if a child receives marks for two spelling tests and for one they get 7 out of 10 and for the next they get 8 out of 10, the marks are often written in common fraction form, $\frac{7}{10}$ and $\frac{8}{10}$; but when the two marks are added the child has 15 out of 20 or $\frac{15}{20}$. In this example, the spelling marks look like fractions and the child thinks that both the numerators *and* denominators are added. This is a very unfortunate misunderstanding and just another reason that fractions can be confusing.

If the fractions we wish to add have different names (denominators), then before we add them a name that suits both is needed (or in other words, a common name, or common denominator). To find a name that suits both fractions, it is important to understand that any common fraction has a limitless number of guises, where the numbers may change but the value does not. Let's look at a scenario to help with this.

## Scenario 7.4

### EATING PIZZA

**Equivalent fractions** are common fractions that have the same value: for example, $\frac{1}{2}$ and $\frac{4}{8}$.

I can eat $\frac{1}{2}$ of a pizza or $\frac{2}{4}$ of a pizza or $\frac{4}{8}$ of a pizza; the fractions may look different, but they refer to the same amount of pizza, as shown in Figure 7.7.

This concept is known as equivalence and the fractions mentioned here are called **equivalent fractions**.

**Figure 7.7** Pizzas cut to show equivalent fractions of $\frac{1}{2}$, $\frac{2}{4}$ and $\frac{4}{8}$

So, to add $\frac{1}{2}$ to $\frac{1}{3}$, a name that suits both must be found (remember that they must have the same name). We have already met some equivalent fractions for $\frac{1}{2}$ in our earlier scenarios. We know that $\frac{1}{2}$ can be $\frac{2}{4}$, $\frac{3}{6}$ or $\frac{4}{8}$ and so on to infinity. Likewise, equivalent fractions for $\frac{1}{3}$ are $\frac{2}{6}$, $\frac{3}{9}$, $\frac{4}{12}$ and so on. If we compare these two sets of equivalent fractions for one-half and one-third, we can now see that $\frac{1}{2}$ and $\frac{1}{3}$ can both be written as sixths – they have a common name in sixths: $\frac{1}{2} = \frac{3}{6}$ and $\frac{1}{3} = \frac{2}{6}$. Therefore, $\frac{1}{2} + \frac{1}{3}$ is the same as $\frac{3}{6} + \frac{2}{6}$ and because these new fractions have the same name, they can be successfully added: $\frac{3}{6} + \frac{2}{6} = \frac{5}{6}$. Returning to our original question, we can say that $\frac{1}{2} + \frac{1}{3} = \frac{5}{6}$.

Again, using the pizza example, the concept of equivalence can be thought about by understanding that the more equal slices a pizza is cut into, then the more slices we will need to eat to have the same amount. For example, if I want to eat half a pizza and it is cut into 12 equal slices, I will need eat 6 slices. Notice that this is a multiplicative relationship: the denominator is six times bigger ($6 \times 2 = 12$) and the numerator is six times bigger ($6 \times 1 = 6$). The lack of understanding of equivalence was exemplified by the famous Yankees baseball player, Yogi Berra, when he said, 'you better cut the pizza into four slices, because I'm not hungry enough to eat six slices.'

## Equivalent fractions

In the previous section we needed to use equivalent fractions in order to ensure that our fractions had the same names before we could add them. Knowing multiplication number facts allows us to find equivalent fractions easily. For example, if we need to add a pair of fractions such as $\frac{1}{3}$ and $\frac{1}{4}$, rather than writing out all the possible equivalent fractions for each to find a common name (denominator), it is quicker to ask, 'what multiple do both denominators have in common?'. (Remember that we discussed multiples and factors in Chapter 4.) We know that 3 and 4 are both factors of 12, so 12 is a common multiple and would be a good denominator $\left( \frac{1}{3} = \frac{4}{12} \text{ and } \frac{1}{4} = \frac{3}{12} \right)$. In this example, the easy way to find the denominator that suits both fractions is to multiply them together ($3 \times 4 = 12$), but this does not always work conveniently. This strategy will give a common denominator but not always the lowest one. Think about $\frac{1}{4}$ and $\frac{3}{8}$; if the common denominator was found by multiplying the two denominators ($4 \times 8 = 32$), the fractions become very difficult to visualise and become unwieldly $\left( \frac{1}{4} = \frac{8}{32} \text{ and } \frac{3}{8} = \frac{12}{32} \right.$, which becomes $\left. \frac{8}{32} + \frac{12}{32} = \frac{20}{32} \right)$. To find the best denominator, go back to the question, 'what multiple do both denominators have in common?' What we really want is the lowest multiple that both denominators have in common. The answer is 8 (8 is a multiple of 4 and 8 is a multiple of 8). The fractions are now much more easily thought about:

$$\frac{1}{4} = \frac{2}{8} \text{ and } \frac{3}{8} = \frac{3}{8} \text{ , so } \frac{2}{8} + \frac{3}{8} = \frac{5}{8}$$

1. Find the lowest common multiple for these denominators:
   a. 5 and 10
   b. 4 and 6

2. Find equivalent fractions for these pairs so they can be added (with the same name or denominator):
   a. $\frac{1}{2}$ and $\frac{1}{6}$
   b. $\frac{1}{3}$ and $\frac{2}{5}$

*(Answers in Appendix.)*

Note that the process for subtraction of fractions is exactly the same as that for adding them. First, we find a common denominator and then subtract the numerators. The name (denominator) stays the same.

## Multiplying fractions

We won't spend too much time discussing the multiplication of fractions; however, it may be helpful to know that it is much easier than addition or subtraction of fractions. This is because the numerators and denominators are multiplied. An example that is easy to visualise is $\frac{1}{2} \times \frac{1}{2}$. Finding $\frac{1}{2} \times \frac{1}{2}$ is the same as asking what is $\frac{1}{2}$ of $\frac{1}{2}$. Think of a pizza again, think of $\frac{1}{2}$ of it, now think of half of that half. The answer is $\frac{1}{4}$, which can be calculated by multiplying the numerators (1 × 1) and the denominators (2 × 2).

The only complication that comes with multiplication of common fractions that you probably remember from school days is the concept of 'cancelling'. Let's look at $\frac{3}{4} \times \frac{4}{5}$ (not something one normally encounters in daily life but let's proceed). First, we multiply numerators together and also the denominators together:

$$\frac{3}{4} \times \frac{4}{5} = \frac{12}{20}$$

Now it's time to think back to your number facts (from Chapter 4). Can you see that 12 and 20 have factors in common? What is the largest common factor? Hopefully you said 4. We can use this knowledge and our knowledge of equivalent fractions to **simplify** $\frac{12}{20}$ by dividing both numerator and denominator by 4 to give us $\frac{3}{5}$ (you may have heard this described as 'cancelling the 4s').

Finding a fraction of whole numbers is useful. Some are linked to number facts: for example, $20 \times \frac{1}{4}$ is the same as $20 \div 4$ (again remember our discussion from Chapter 4 when we looked at related or derived facts). Some calculations can go beyond number facts: for example, $72 \times \frac{3}{4}$ is more complicated. In this case we find $\frac{1}{4}$ of 72 (or $72 \div 4 = 18$) and then multiply 18 by 3 to get 54.

## Dividing fractions

While we will not spend much time on division of fractions, some simple examples may be encountered in our everyday lives. For example, if I have one-half metre of ribbon and I need to cut it into equal pieces that are one-eighth metre for a craft project, how many pieces can I make? We can represent this problem as $\frac{1}{2} \div \frac{1}{8}$. We can use our number sense to solve this question because we know that $\frac{1}{2}$ is the same as $\frac{4}{8}$. This means that if I divide my one-half metre of ribbon into one-eighth pieces, I will get four of them.

The great news is that you don't have to formally do a division by common fractions. Think of the simple number fact $8 \div 2 = 4$; now think of $8 \times \frac{1}{2} = 4$. These two equations have the same starting number and the same answer, which logically means that the instructions '÷ 2' and '× $\frac{1}{2}$' must perform the same function. This idea is called reciprocity and the numbers 2 and $\frac{1}{2}$ are known as **reciprocals** of one another. So, when a division is encountered the reciprocal can be used instead. For example, '÷ 4' is the same as '× $\frac{1}{4}$', and '÷ $\frac{1}{3}$' is the same as '× 3'.

---

## Learning activity 7.7

1. Calculate (and simplify if possible):

    a. $\frac{5}{8} \times \frac{1}{2} =$

    b. $\frac{3}{4} \times \frac{1}{3} =$

2. Calculate:

    a. $25 \times \frac{1}{5} =$

→

**b.** $\dfrac{2}{3} \times 24 =$

**3.** Calculate:

**a.** $3 \div \dfrac{1}{6} =$

**b.** $4 \div \dfrac{1}{5} =$

*(Answers in Appendix.)*

# SOME POSSIBLE REASONS FOR DIFFICULTY WITH COMMON FRACTIONS

Many people report a history of difficulties with fractions, particularly common fractions. At university many pre-service teachers will admit that working with fractions is a weak point. It is quite interesting that many students can still recall quite vividly the moment or grade or teacher when difficulties with fractions first developed.

Young learners spend their initial years at school mainly focused on whole-number thinking, as they learn to add, subtract, multiply and divide and concentrate on working with their number facts. The step to fractional thinking requires a considerable increase in complexity, and sometimes the entrenched whole-number thinking can be an obstacle rather than a stepping stone.

Some education researchers have described the movement into fractional thinking as a child's first experience of abstraction in mathematics (Booker et al., 2014). If the step into this abstraction is attempted too early, too quickly or too procedurally, then difficulties are likely to arise. For a child who experiences any of these, it is often the beginning of a spiralling loss of confidence, which effects their general numeracy and number sense; this often persists into adulthood. If you still feel that fractional thinking is an issue, consider the following list of fractional concepts. Which do you feel are now clarified? Which are still challenging you?

- When referring to common fractions, such as $\dfrac{2}{3}$, it must be remembered that the parts (in this case, the thirds) must be equal in size. Children generally understand this because it relates to fairness: try giving a child the 'small' half of a piece of cake!

- A proper common fraction (a part of a whole) has a value between 0 and 1. Some people think that a fraction is less than zero.

- When comparing fractions with equal numerators, such as $\dfrac{1}{2}$ and $\dfrac{1}{3}$ or $\dfrac{2}{5}$ and $\dfrac{2}{6}$, the fraction with the larger denominator is the smallest, so $\dfrac{1}{3}$ is less than $\dfrac{1}{2}$ (think of a pizza again). This is tricky for children when they first encounter common fractions because they are used to whole-number thinking where 3 is bigger than 2.

- When multiplying by a proper fraction (or finding a fraction of something), the answer is less than the number we started with: for example, $10 \times \dfrac{1}{2} = 5$. Again, this is contrary to what young learners originally experience where multiplying by whole numbers makes things bigger.

- Inversely, dividing by a fraction gives an answer larger than the starting number: for example, $4 \div \dfrac{1}{2} = 8$. Rather than getting confused by this it is easier to ask the question another way: How many times does $\dfrac{1}{2}$ go into 4? (Again, thinking about a pizza might help – how many half-pizzas are in four whole pizzas?)

- Some fractions we may have worked with in school are difficult to imagine (more difficult than whole numbers): for example, $\dfrac{13}{17}$. These kinds of fractions are fine for working purely in the mathematical world, but it is unlikely that they will be encountered in the real world. People with highly developed number sense might still struggle with this, so do not fret if working with complex fractions still causes difficulties.

# DECIMALS

We have dealt with many of the concepts around decimals in Chapter 3, when we discussed place value. Decimals or decimal fractions (various countries refer to them in different ways) require fractional thinking and have similarities and differences with common fractions. Just as with common fractions, decimals refer to numbers between 0 and 1. Strangely, some people think that decimals and common fractions aren't numbers, but they are, and they have their own place in the number system; as with common fractions, using a number line helps to see this.

**Figure 7.8**   Decimal fractions lie on the number line between 0 and 1

First and foremost, to understand decimals it is essential to have a clear understanding of how our Hindu-Arabic number system works, so it may be helpful to review Chapter 3 at this time. As we discussed in Chapter 3, at the core of understanding decimals is knowing about Base 10, place value and the use of zero. After the whole numbers and the decimal point, the decimal fractions begin with tenths, then hundredths and then thousandths, and so on. In our daily lives, it is rare that we need to go beyond thousandths. For example, when dealing with money we use two decimal places; when we use the bathroom scales, most use one decimal place.

The place value (column) names of tenths, hundredths and thousandths do the same job as the denominator in common fractions. So, in a number like 0.4 (said 'four-tenths'), the first number (4) is the numerator and the second number (10) is the denominator, which is the same as $\frac{4}{10}$. Herein lies a very important difference between common fractions and decimal fractions: the common fractions can have any number for a denominator, but decimals are restricted to the place value columns of tenths, hundredths and so on. These different attributes give each type of fraction some different strengths and uses. The common fractions can be incredibly specific in their denominators (e.g. $\frac{1}{512}$ or $\frac{1}{893}$) but such fractions are not often used in day-to-day life. The decimals, while having a set range of denominators, are much more regularly used in real life, simply because they are linked to the Hindu-Arabic number system and are used in all our electronic devices (i.e. calculators and computers).

Just as with common fractions, learners have to overcome their initial focus on whole-number thinking and realise that hundredths are smaller than tenths of the same object or collection. Fortunately, our early understanding of decimals is enhanced through real-life experiences with money (dollars and cents); though according to many primary school teachers we work with, this is becoming less so, as the use of non-cash payment methods increases.

## Working with decimals

Most of our students who report difficulties with fractions are referring to common fractions. They are generally more comfortable with decimals mainly because of their easy application to calculators. However, there are a couple of 'tricks' to be aware of.

As decimals are a continuation of the Hindu-Arabic number system, place value is of ultimate importance when calculating; any errors in the location of numbers will lead to an incorrect answer. Also remember that the decimal point does not move, even though you may hear people say this. Recall as well the position of the decimal point we discussed in Chapter 3. The decimal point signifies the end of the whole numbers and sits in the ones column. The numerals move left or right across the decimal point when calculating. Look at Figure 7.9. You can see that when the number 3.15 is multiplied by 10, it becomes 31.5 and when it is divided by 10, it becomes 0.315. Note that the decimal point remains in place and the digits move left in the case of multiplication and right in the case of division.

| Tens | Ones. | tenths | hundredths | thousandths |
|------|-------|--------|------------|-------------|
|      | 3.    | 1      | 5          |             |
| 3    | 1.    | 5      |            |             |
|      |       |        |            |             |
|      | 3.    | 1      | 5          |             |
|      | 0.    | 3      | 1          | 5           |

$3.15 \times 10 = 31.5$

$3.15 \div 10 = 0.315$

**Figure 7.9** Multiplication and division by 10 shift the digits in place values, not the decimal point

# Learning activity 7.8

Calculate the following (take note of how the number 'slides' left or right across the decimal point):

1. $51.6 \times 10$
2. $23 \div 10$
3. $0.4 \times 10$
4. $7 \div 10$

   *(Answers in Appendix.)*

Important decimal concepts can be drawn out in the following example. Children notoriously have trouble with this type of question: Which is the larger number, 0.1 or 0.09? Again, it is their whole-number thinking that confuses them; they see the 9 and assume it is bigger than the 1. However, remember that it is the place that gives value (place value), so 0.1 (tenths) is bigger than 0.09 (hundredths). A way of thinking about this is to put a zero after the 1, making it 0.10; this does not change the value but perhaps allows a more visual comparison: 0.10 and 0.09. Now both numbers are seen as hundredths and are more easily compared. Note: placing zeros after decimals does not change the value and is usually a pointless exercise, but it helps in the above example. Similarly, placing zeros before whole numbers does not change the value.

# Learning activity 7.9

Write the following groups of numbers in ascending order (i.e. from smallest to largest):

1. 0.8, 0.08, 0.18
2. 0.1, 0.04, 0.01

   *(Answers in Appendix.)*

Another question worth considering because it requires deep understanding of decimal places and our number sense is: What number is halfway between 0.3 and 0.4? This causes some students problems but if the question is rephrased into the hundredths, then again, the answer becomes more obvious: What number is halfway between 0.30 and 0.40? Responses are even quicker if the question is: What amount is halfway between $0.30 and $0.40?

## Learning activity 7.10

Find the number that is halfway between:

1. 0 and 1
2. 0.2 and 0.4
3. 0.5 and 0.6
4. 0.03 and 0.04

   *(Answers in Appendix.)*

# PERCENTAGES

Percentages are another type of fraction and are very much a part of everyday life. Having good number sense with percentages can lead to saving lots of money, as they are commonly encountered when spending it. For example, when shopping we might be interested in discounts. When we are looking for a loan, we would be focused on the interest rate. The term 'percent' is represented by the symbol % and means per hundred from the Latin 'centum'. The often overlooked part of the terminology is the 'per'. We use 'per' commonly in phrases such as 'kilometres per hour' where it is part of an expression of rate (covered further in the next chapter). Percentages are ultimately an expression of a rate. Because all percentages are per 100, this allows percentages to be used as a fraction of comparison.

## Scenario 7.5

## RESTAURANT REVIEWS

A restaurant's owners want to ensure the restaurant is constantly trying to improve to meet customers' expectations. In one week, 35 people out of 83 gave the restaurant a five-star review and in the following week 51 out of 139 gave it five stars. These are awkward figures to compare (imagine trying to compare $\frac{35}{83}$ with $\frac{51}{139}$). If the owners are absolute thinkers (we discussed absolute thinking in Chapter 6), then they would be happy that the number of five-star reviews has gone up from 35 to 51, but if they are relative thinkers then they might use percentages to compare the ratings for the two weeks. In the first week, approximately 42% (35 ÷ 83 × 100) of customers gave a five-star review and in the second week it was approximately 37% (51 ÷ 139 × 100) of customers. So, the percentages have given a clear picture that there was an actual decline in five-star satisfaction because they allow a true (relative) comparison.

Notice that in Scenario 7.5, to find the percentages the number of five-star reviews was divided by the total number of customers (giving a decimal answer) and then multiplied by 100 to give the percentage; try this on a calculator to see it in action.

---

## Learning activity 7.11

Convert the following pairs of fractions into percentages and identify the larger in each case:

1. $\dfrac{7}{8}$ and $\dfrac{16}{18}$

2. $\dfrac{107}{300}$ and $\dfrac{46}{139}$

   *(Answers in Appendix.)*

---

Percentages, like common fractions and decimals, can be thought of as having a numerator and a denominator. Common fractions have limitless denominators, decimals have set denominators (tenths, hundredths, thousandths and so on) and percentages have just one denominator, which is 100. This might be seen as a disadvantage, but it is what allows comparison because all expressions of percentages have the same denominator (name).

While the restaurant example above had quite complicated numbers, and needed a calculator, most of our everyday life encounters with percentages can be worked out quite simply in our heads. The common theme continues, in that to have functional number sense with percentages there is a need to understand the Hindu-Arabic number system and know your number facts. We most often encounter percentages when shopping and mostly when being attracted by discounts. Because percentages mean per 'hundred' they fit beautifully with the Base 10 nature of our number system and the key to dealing with percentages in our heads as we're wandering through the shop looking at sale items is to consider 10%, which also fits with our Base 10 thinking.

Knowing 10% of anything is simply a matter of sliding the number one place to the right (10% = $\frac{1}{10}$, which means the number is getting smaller and so we move the number to the right). For example, 10% of 1 243 is 124.3, 10% of $59 is $5.90, 10% of 480 km is 48 km. Because 10% is easy to find, then multiples of 10% can be derived. For example, 20% is twice 10%, 40% is 4 times 10%. This is the strategy to use when shopping.

The idea behind finding 10% can be expanded to accommodate other discounts. For example, if there is a 5% discount we can find 10% and then halve it. To find 5% of $120, think: 10% of $120 is $12, so 5% is half of $12 = $6. Sometimes finding 1% of something is also useful and to do this we employ a similar strategy to the 10% method. Because 1% = $\frac{1}{100}$, we are now looking for a number 100 times smaller, so we slide the number two places to the right. For example, 1% of $750 = $7.50.

Other handy mental computations with percentages are the quite common discount amounts of 25%, 50% or 75%. For each of these it is easier to think of their common fraction equivalents (25% is $\frac{25}{100}$ or $\frac{1}{4}$) so to find the 25% discount amount, simply divide the price by 4 (e.g. 25% of $200 is the same as $\frac{1}{4}$ of $200, which is the same as $200 ÷ 4). We know that 50% is the same as one half, so we can divide the price by two; and 75% is the same as $\frac{3}{4}$, so find $\frac{1}{4}$ (same as 25%) and then multiply by 3.

Calculate each of the following amounts (try to use the above strategies and mental computation to do this):

1.  The amount saved on an item priced at $35 with a 10% discount
2.  The amount saved on an item priced at $60 with a 30% discount
3.  The sale price of an item priced at $70 with a 40% discount
4.  1% of 1 435
5.  5% of $45
6.  50% of 640 km
7.  75% of $240

    *(Answers in Appendix.)*

While the methods described in this section can help in most day-to-day situations involving percentages, the most commonly reported difficulty is interpreting the small print on the discount signs. Here are some to consider:

*   25% off lowest marked price
*   25% off RRP (recommended retail price)
*   25% off selected items
*   Up to 25% off storewide
*   25% off second item of equal or lower price
*   Take a further 25% off already discounted items

**Figure 7.10**   Different sale discount 'deals'

Although we haven't delved into the use of percentages in the financial world, it is worth mentioning that when you are advised to read the fine print on any financial/legal documents, it is extremely important that you do so. While the percentages mentioned in the documents might look small, the wording around percentages can have a devastating effect. Think about people who get into credit card debt and cannot pay off the interest so get further and further into debt; think about people who draw down on their mortgage to fund an item or trip and the interest they will be paying on it for the life of the mortgage (perhaps 25 years). The term 'debt trap' is very apt in these situations. Having sound number sense about percentages and an understanding of the associated language can impact one's life enormously, both positively and negatively.

## WORKING ACROSS THE THREE MAIN FRACTION TYPES

While it is important to understand each of the fraction types discussed in this chapter, it is also very important for our number sense to be able to move seamlessly from one type to another. Walking around a grocery shop you will encounter fractional expressions mainly in common fractions and percentages for discounts, and of course lots of decimals in measures and dollars and cents. Thinking of fractions as numbers on a number line can help us see equivalences: for example, $\frac{1}{4} = 0.25 = 25\%$ or $\frac{1}{5} = 0.2 = 20\%$.

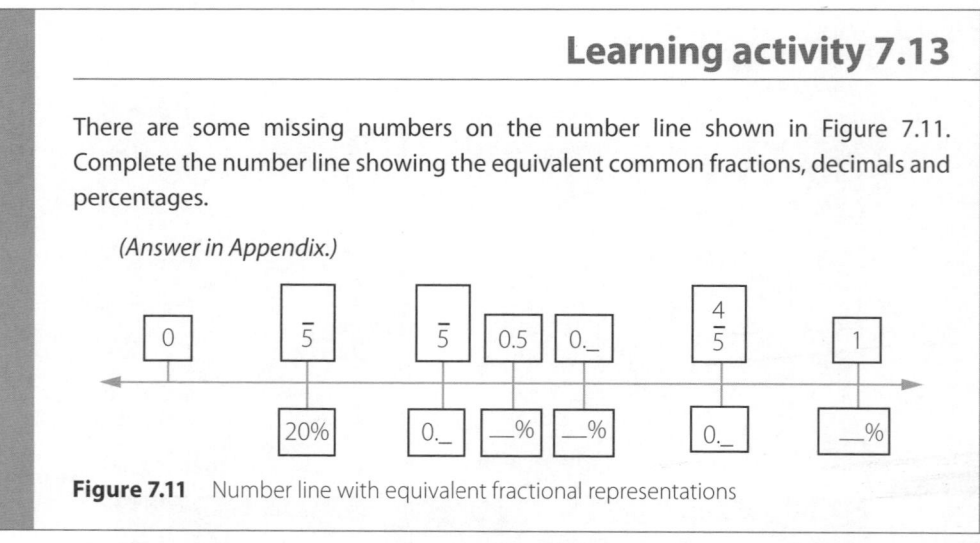

### Learning activity 7.13

There are some missing numbers on the number line shown in Figure 7.11. Complete the number line showing the equivalent common fractions, decimals and percentages.

*(Answer in Appendix.)*

**Figure 7.11**  Number line with equivalent fractional representations

Exploring with a calculator may help you understand how the fractions link. Think of the common fraction $\frac{5}{8}$. To put this on a calculator press 5 ÷ 8 = (the vinculum acts as a division); you will get a decimal answer, 0.625; now multiply that by 100 and you will get the percentage 62.5% (although the % sign may not show).

The data in Figure 7.12 are taken from a display in the Titanic Museum in Belfast.

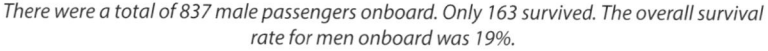

*There were a total of 837 male passengers onboard. Only 163 survived. The overall survival rate for men onboard was 19%.*

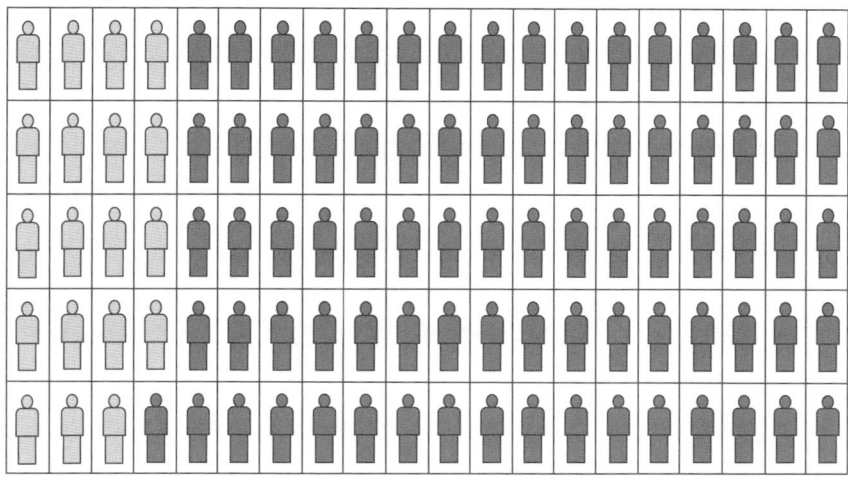

81% of men died

**Figure 7.12** *Titanic* shipwreck data showing survival rate for male passengers

Notice the different representations used in Figure 7.12 to show the same information in different ways. We will come back to multiple representations in the chapter on problem solving.

# Learning activity 7.14

Look at the different ways that the information has been displayed in Figure 7.12.

1. Why do you think the survival rate was presented as a percentage?
2. What is the purpose of the visual representation?
3. Would it have been as meaningful if the display had said ' $\frac{674}{837}$ male passengers died' instead of telling you the percentage (81% died)?
4. Use your calculator and the steps presented in the paragraph before the figure to calculate $\frac{163}{837}$ as a decimal fraction.
5. By converting the decimal to a percentage, verify that the percentage shown in the figure is correct.

*(Answers in Appendix.)*

# CONCLUSION

In this chapter we have looked at some of the fundamental ideas needed in fractional thinking. We focused in some detail on common fractions because these are known to be challenging for many people. We looked at their structure, the ways they can be used and interpreted and briefly at calculating with them. The chapter also focused on decimals and percentages because they too involve and support fractional thinking. Finally, we described the links among common fractions, decimal fractions and percentages. The ideas in this chapter contribute to number sense because they deepen our understanding of relationships among numbers and representations and help us to understand operations and relative size.

## Personal actions to improve number sense

By following Polya's problem-solving strategy, it is possible to now *carry out the plan* to improve your personal number sense. The following are suggestions to assist you:

- Identify the three fraction types in your day-to-day life (i.e. common fractions, decimals and percentages).

- If a fraction appears in one form, consider how it would look in a different form (e.g. 30% discount would be $\frac{3}{10}$ or 0.3).

- If you have a mortgage or loan, check out the percentages being paid in interest; maybe you could renegotiate to find a lower percentage and save some money. Remember, a small percentage difference over a long time can make a substantial dollar difference.

- When out shopping, deliberately check out the language associated with discounts and not just the discount fractions.

- Employ the 10% strategy for calculating discounts, even if you are not buying.

# 8

# Ratio, rate and scale

## LEARNING OBJECTIVES

After reading this chapter, you should be able to:

- understand and apply ideas related to ratio
- understand and apply ideas related to rate
- make connections between ratio and rate
- understand some applications of scale in everyday situations
- connect concepts of multiplicative and relative thinking with ratio, rate and scale.

# INTRODUCTION

Throughout this book we have worked through a number of main elements involved in developing number sense. In this chapter we focus on ratio, rate and scale, which encompass some of the key circumstances that allow the application of the knowledge and strategies of number sense we discussed in the previous chapters. While the three topics are inextricably linked, they will be dealt with in turn, but with no doubt some crossover of concepts. These topics rely heavily on relative thinking and multiplicative thinking, discussed in Chapter 6. The topics, particularly ratio and rate, can be challenging for students and indeed for many pre-service teachers (Livy & Herbert, 2013). A study conducted in 17 countries by Tatto et al. (2012) found these topics to be problematic for both primary and secondary pre-service teachers. It is quite challenging to write clear, accurate and succinct definitions of the terms 'ratio', 'rate' and 'scale', and we shall discuss this as we proceed, keeping in mind that our focus is on number sense and not on all the variations and applications of these ideas. In this chapter we will use familiar examples to support your understanding of the concepts that underpin them. The chapter begins with ratio.

# RATIO

A **ratio** is a comparison between quantities with the same units of measure: for example, 3 cats to 4 dogs or 3:4.

A quick Google search on the meaning of the term **ratio** will convince you that there are multiple definitions, and this is likely to be a problem for anyone who is uncertain or lacking in confidence when it comes to these ideas. The definitions vary and include 'a comparison of quantities' and 'the number of times one number is contained in another'. These are not particularly helpful without further explanation or examples. In this chapter we will use the definition from the Australian Curriculum: A ratio represents a comparative situation – it is a comparison between quantities with the same units of measure (ACARA, 2020b). When we refer to units of measure we mean both parts of the ratio deal with animals or people or plants or liquids, and so on. For example, if we read that the ratio of teachers to students in Australian government schools is 1:14, this means that for every 1 teacher there are 14 students (this ratio deals with the common measure of people).

As we proceed through the chapter, the definition and application of ratio will be expanded through examples. In the next section we will look at some of the basics of ratio and common representations.

## Representing ratio

As you would have noticed from the margin definition, ratios are mathematical expressions that use the symbol ':' between the quantities being compared. For example, the ratio of dark-leafed lettuce to light-leafed lettuce in Figure 8.1 is 2:3. This means that for every 2 dark-leafed lettuce there are 3 light-leafed lettuce.

**Figure 8.1**   The ratio of dark-leafed lettuce to light-leafed lettuce is 2:3

The ratio in Figure 8.1 is read as '2 is to 3' and represents the relationship between the two parts. When expressing relationships using ratios, the order matters. For example, in Figure 8.1 the ratio 2:3 represents dark-leafed lettuce to light-leafed lettuce. On the other hand, if we write 3:2 we are expressing the ratio of light-leafed lettuce to dark-leafed lettuce. We could express the ratio of dark-leafed lettuce to the whole group as 2:5, which means that for every 5 lettuce, 2 are dark-leafed. We can also express this ratio using fractions (as discussed in Chapter 7). We could say that 3 out of every 5 lettuce are light-leafed, so $\frac{3}{5}$ of the lettuce are light-leafed. Notice that when we use fractions in this way, we are comparing one part to the whole collection.

Ratios exist in a multiplicative relationship, and as with common fractions we can have equivalent ratios: for example, 3:2 is the same ratio as 6:4 or 9:6. Also, just like common fractions, ratios can be (and usually are) written in lowest terms, so a ratio of 6:8 would be reduced to 3:4 by dividing the two elements of the ratio by 2. Percentages also express ratios. For example, 30% means the same as 30:100. These relationships between ratio and percentages or fractions can be useful in real life because we are more familiar with fractions and percentages and it is often the case that we don't speak using the language of ratio. For example, if a bag of sweets has 12 red sweets and 8 blue sweets, the ratio of red to blue is 12:8 (or 3:2 in lowest terms) but we probably wouldn't use this in everyday speech. Instead, we might say that 12 of the 20 sweets are red ($\frac{12}{20}$ or $\frac{3}{5}$ in lowest terms) and 8 of 20 sweets are blue ($\frac{8}{20}$ or $\frac{2}{5}$).

## Learning activity 8.1

For each of the following examples, complete the sentences to write the relationship as a ratio in lowest terms (using :) and as a fraction (remember that expressing as a fraction requires you to consider the whole, not just the parts).   →

Now that we have revised the basics of ratio, we will look at the development of ratio and its use in the real world.

## Developing ratio understanding

When children first learn to count to 10, they will be very proud of themselves when they can say all 10 digits (in order!). You may have noticed, however, that they cannot always successfully count objects. This is because they have not yet developed one-to-one correspondence, where each successive object is given the next number in the sequence. Developing one-to-one correspondence is a major step for young children as, among other things, it is the most basic of ratios; each object being counted has a number assigned to it: 1 object to 1 number. The photo in Figure 8.2 represents a 1 to 1 correspondence because there is one post cap for each fence post (i.e. the ratio of post caps : fence posts = 1:1).

**Figure 8.2**    A farm fence with one post cap for every one fence post

A step forward comes when children start activities, such as graphing, especially pictographs. Initially, the graphs will emphasise one-to-one correspondence – for each car counted in the car park, one car is drawn on the pictograph. The next step is to represent the cars using a ratio (a legend in the graph) of perhaps 2 real cars represented by 1 drawn car on the graph. Scenario 8.1 represents such an example.

## Scenario 8.1

## CARS IN THE PARKING LOT

The Year 3 children in Mrs Brown's class were asked to count the number of teachers' cars in the parking lot and represent their data using a pictograph, with 1 car on the pictograph representing 2 cars in the parking lot. The children's work is shown in Figure 8.3.

| Car colour | Number of cars |
|---|---|
| Red | |
| Blue | |
| White | |
| Black | |
| Green | |

 Legend = 2 cars

**Figure 8.3** Pictograph showing cars with 1:2 ratio

It is at this point that children begin working with ratio, before engaging in formal calculations with them. This kind of activity can again be aligned with activities like skip counting, repeated addition, and number facts and ultimately multiplicative thinking and ratio. So again, these ideas are a powerful stepping stone in the development of number sense.

Imagine that the pictograph in Scenario 8.1 represents the number of cars in the local shopping centre parking lot and that the car picture in the legend now represents *10* actual cars.

1. How many cars in the parking lot are red?
2. How many cars in the parking lot are blue?
3. What is the ratio of green cars to black cars in the parking lot?

   *(Answers in Appendix.)*

## Real-world examples of ratio

If the previous learning activities felt a little too much like the school mathematics you found daunting, remember that growth mindset! In this section we will look at some real-world uses of ratios. The most important part of understanding ratio in everyday life is knowing that the multiplicative relationship is maintained as circumstances require quantities to grow or contract: for example, making a larger or smaller amount of a recipe or mixed drink (as we did in Chapter 6). When we need to maintain the relationship, we refer to this as keeping things in proportion. For example, if my raspberry cordial label tells me to mix 1 part cordial for 5 parts water, I know that this means I will need to keep this ratio when I make my drink. If I am using 500 mL of water, I will need 100 mL of cordial or if I am using 1 000 mL of water, I will need 200 mL of cordial (because 1:5 = 100:500 = 200: 1 000 and I need to maintain the ratio). Keeping ingredients in proportion is essential for maintaining the same flavour.

## Learning activity 8.3

We recently saw the sign in Figure 8.4 on the side of a blood bank vehicle. What do you think is the main message the sign is intended to convey by using ratios?

*(Answer in Appendix.)*

**Figure 8.4**  Blood bank vehicle signage

---

Look at the information in Table 8.1.

**Table 8.1**  Nurse/midwife ratios and infant mortality in global regions

| Region | Northern America | Latin America | Europe | Africa | Asia |
|---|---|---|---|---|---|
| Nursing and midwifery personnel compared to population | 1 midwife/ nurse for every 100 people (1:100) | 1 midwife/ nurse for every 282 people (1:282) | 1 midwife/ nurse for every 121 people (1:121) | 1 midwife/ nurse for every 884 people (1:884) | 1 midwife/ nurse for every 489 people (1:489) |
| Infant mortality | 6 children in every 1 000 (6:1 000) | 17 children in every 1 000 (17:1 000) | 5 children in every 1 000 (5:1 000) | 51 children in every 1 000 (51:1 000) | 28 children in every 1 000 (28:1 000) |

Source: Adapted from Population Connection (2016).

What can you say about the relationship between the ratio of nurses/midwives in a region and the region's infant mortality ratio?

*(Answer in Appendix.)*

As adults, we probably most often encounter ratios in our daily lives in various contexts such as cooking, taking medications or gardening. Recipes are essentially a mix of numerous ingredients, all of which must be kept in proportion to maintain the desired characteristics (e.g. taste of food, effectiveness or safe dosage of medication, concentration of chemicals).

## FERTILISING THE VEGETABLE PATCH

I want to apply some liquid fertiliser to my vegetable patch and my indoor pot plants.

The label on the fertiliser bottle says that 10 mL of fertiliser should be mixed with 3 L of water. I have three different watering cans. The largest holds 9 L, the medium-size metal watering can holds 2 L and the smallest can holds 1.5 L.

### Learning activity 8.5

I need to use the instructions in Scenario 8.2 (i.e. 10 mL : 3 L) to measure the correct amount of fertiliser.

How many millilitres of fertiliser would be needed for:

1. the 9 L watering can?
2. the 1.5 L watering can?
3. the 2 L watering can?

My largest and smallest watering cans are easy to calculate but I always have difficulty with the 2 L can. Can you see why this might be?

*(Answers in Appendix.)*

Note that in Scenario 8.2 and Learning activity 8.5 the ratio of fertiliser to water was expressed as 10 mL : 3 L. This is not strictly mathematically correct because we usually convert the quantities in the ratio so that they are in the same units. For example, in this scenario 10 mL : 3 L would be converted to 10:3 000 or 1:300. In practical situations, however, it is easier to understand and use the ratio as it is expressed on the fertiliser bottle (10 mL : 3 L).

Some situations in life do not require such strict adherence to calculated ratios and instead rely on approximations and, as we shall see in Scenario 8.3, are sometimes even related to instinct or gut feeling.

## Scenario 8.3

## SAFETY IN NUMBERS

Have you ever been in a situation where you have felt safer with a group of people (safety in numbers)? The feeling of safety comes because rather than facing a perceived danger or threat by yourself, you increase the ratio of companions to threat in your favour by joining with friends, or even by blending into a crowd. Interestingly, this situation was the opposite in the days of the COVID-19 pandemic when it was safer to be in non-crowded spaces.

Animals can also have an instinctive understanding of ratios. During the annual migration season of grazing animals in East and Southern Africa, at some point the animals reach a river crossing, which is crocodile infested. Early in the morning a few zebras and wildebeest arrive at the riverbank but do not cross. The ratio of migrating animals to crocodiles at this point is 'few : lots'. As the day progresses, the number of zebras and wildebeest on the bank grows and grows into the thousands, with associated dust and noise. The ratio of migrating animals to crocodiles has now changed to 'thousands : relatively few'. At this point one brave creature enters the river and all the rest follow in an amazing display. Yes, some are taken by crocodiles, but most survive. The zebra and wildebeest have waited for safety in numbers.

1.  Describe other situations where you have changed your behaviour according to ratios or alternatively changed a ratio to suit your behaviour.
2.  Can you think of other situations in nature that show the benefits of a favourable ratio?

The next section is focused on rate and builds on the ratio concepts introduced in this section and the relationships involving relative and multiplicative thinking we looked at in Chapter 6.

# RATE

**Rate** is a ratio comparing two different types of quantities, which usually have different units.

As with ratio, 'rate' is variously defined. In this chapter we define **rate** as a ratio comparing two different types of quantities, which usually have different units (ACARA, 2020b; De Klerk, 2014). Everyday examples where rate is used include dollars per litre of fuel (buying fuel), kilometres per hour (speed), alcoholic units per day (health warnings) or kilojoules per day (nutrition).

Rate is one of the most important concepts to understand in our daily lives, because often it is not the activity we undertake (good or bad; healthy or unhealthy) but the rate at which we undertake it that affects our lives, either positively or negatively. Sometimes a rate can seem inconsequential but if an activity or life habit is done regularly over a substantial period of time, it can still have a substantial impact. For example, a tap dripping at one drop per minute can eventually flood a room; a small and regular rate of saving money can result in considerable wealth. In the next section we look at some commonly encountered rates.

## Common rates in everyday life

Rates can refer to formal situations and less formal. Formal rates might be used by corporations to monitor production; for example, iron ore producers monitor how many tonnes of ore are shipped per day, week or month so they can track variations in productivity; agricultural producers measure harvest per hectare for the same reason. Rates might be used by companies or shareholders to monitor profit or return on investment. These examples show that rates can be used as a measurement tool and to help inform decisions. In our own lives there are many rates that are less formal, such as how many cans of dog food our pet eats in a week, but understanding these rates can still help inform our decision making.

## Rates of consumption

Knowing how quickly we consume certain items can help us make plans about their replacement (e.g. formulating a weekly shopping list). We might need to think ahead about the amount of a certain item we have left and estimate how much we need to buy (e.g. milk, bread, paper towel).

## Using a car

Another context in which we use rate quite often is when using our cars. For example, modern cars indicate their rate of fuel consumption – usually in litres per 100 km. Have you ever noticed how this rate changes when you drive on the highway and then drive through a town or city?

Perhaps one of the most commonly encountered rates when driving is speed; this is expressed as how many kilometres we travel in an hour (km/h). We will use this example to investigate some important ideas. Speed when driving can refer to a number of things: current speed, average speed for the journey, or the speed limit or advisory speed for a particular section of road. The telemetry of some modern cars goes even further by calculating the average speed of the car for its entire working life. Each of these speed calculations is an expression of the relationship between distance and time. The average speed of a car that travels 100 km in 2 hours is literally 100 km per 2 hours (100 km / 2 h) but this is nearly always expressed as a 'per hour' rate. So, both the distance and time need to be halved, making the speed 50 km/h. Notice the way the rate is expressed: it is like a fraction with a vinculum, which indicates the division that is occurring. The 'current speed' of a car, as indicated on the speedometer, is a 'moment in time' calculation of distance being travelled in that time and this is why it can change upwards rapidly when accelerating or downwards rapidly when braking.

The 'speed limit' or 'advisory speed', which we see on road signs, indicates to the driver a calculation that road authorities have deemed is safe, again a calculation involving distance and time.

New technology that calculates average speeds can be seen on highways; the cameras identify the vehicle and measure the time it takes it to travel over a set distance. From this example, it can be seen that there is a multiplicative relationship (or a divisional relationship) between two factors that inform us of a rate. The relationship between distance and time affords a particular name: speed; but not every rate has a specific name.

---

## Learning activity 8.7

1. I travel 300 km in 4 hours. What is my average speed?
2. I need to travel 150 km. I know that during rush hour the average speed I can expect to drive is 50 km/hour. How long is this journey likely to take me?

*(Answers in Appendix.)*

The road sign in this photo is located on a narrow and winding stretch of road in England.

What is the average rate of accidents per year?

*(Answer in Appendix.)*

## Spending money

Another context in which we commonly encounter rate is when we are spending money. Australian grocery stores have been listing the unit price for items they sell since 2009. Scenario 8.5 illustrates the use of rate for describing the costs of fruit and vegetables.

## Scenario 8.5

## SHOPPING AT THE GREENGROCER

The photos in Figure 8.5 show a number of sale items listed on the wall of our local grocery store. We found this interesting because several were expressed differently in terms of the quantities offered for sale.

**Figure 8.5** Prices of fruit and vegetables

Look at the rates indicated in Figure 8.5. In the first row of three signs, the prices are all given as rates in $ per kg ($/kg) but in the second row of four signs, the rates are all different, which requires us to think differently than for the first three. Why might these varying rates be used?

*(Answer in Appendix.)*

Rates are also important when we are making decisions about spending money in other contexts. For example, comparing interest rates is an essential aspect of choosing a personal loan, credit card or mortgage. Having good number sense when it comes to these rates can help make important life decisions.

## Long-term effects of rates

There are some rates in our lives that help us enormously if they are effectively monitored. Like a number of concepts addressed in this text, our use of rate often does not entail strict calculations. With rate it is important to have an overall understanding of the concept and how it can affect our life decisions and hence our lives. Many of our behaviours will have minimal impact on our lives if they are just 'one-offs'; however, if they occur more regularly then they can have amazing positive or negative effects. Exercising for a day will have little impact on our lives but regular exercise (e.g. at a rate of 30 minutes per day for 4 days per week) has a wonderful positive result. The opposite occurs with our consumption of junk food: a one-off indulgence may have limited long-term effects, but regular consumption (e.g. 5 nights/week) is detrimental. The important factor here is not the behaviour but the rate at which we engage in the behaviour. We need to understand what an acceptable rate for exercise or junk food consumption might be. There are guidelines about rate that are available for activities such as exercise (number of hours per week, number of days per week), junk food consumption (some foods are designated a 'sometimes food', indicating low rate) or healthy daily water consumption.

While some of the above examples may not require strict mathematical monitoring, let's consider some calculated examples of situations that perhaps would not normally be thought about through the notion of rate.

## Scenario 8.6

## THE DAILY COFFEE

Mack visits a local café with her friends every morning. She believes that a cup of coffee and cake at a café is one of life's simple pleasures, and at around $10 per visit it is relatively inexpensive. But in this scenario it is the number of days per week that Mack indulges on that should perhaps be considered. If the coffee and cake is a daily habit, then that is a spending rate of $10 per day ($10/day). If Mack actually does this every day of the year, that amounts to 365 coffees and cakes per year, which would cost 365 × $10 = $3 650. If this is something done throughout a 40-year working life, $3 650 per year becomes $146 000 in today's dollars!

If daily coffee and cake is considered a priority, then that is perfectly fine. It is obviously a personal decision, but as discussed in the next chapter on problem solving, knowing all the facts is important. Knowing the cumulative cost of this habit is crucial so decisions can be made. Monitoring the weekly rate of consumption can result in behavioural changes: for example, indulging in coffee and cake twice per week only.

## Learning activity 8.10

Buying a car is an important decision. One thing to consider is the fuel consumption, which is expressed as a rate of litres of fuel per 100 km. Besides being an environmental decision, buying a vehicle that uses less fuel can also be a huge financial decision. In this learning activity we compare the cost of driving the two cars.

Car A has an average fuel consumption of 8 L per 100 km.

Car B has an average fuel consumption of 5 L per 100 km.

1.  On a road trip of 300 km, how many litres of fuel will each car use? (this is multiplicative)

→

2. If the cost of fuel is $1.50/L, how much will it cost for each car? (this is multiplicative)
3. How much does the owner of Car B save compared to the owner of Car A? (i.e. how much less does it cost for Car B's road trip?) (this is additive)

Now let's consider the total fuel cost over the life of the two cars. We will assume that each car drives 300 000 km over its lifetime and that the average fuel cost is $1.50/L.

4. How much will the owner of Car B save over the life of the car compared to the owner of Car A? (this is multiplicative and additive)

*(Answers in Appendix.)*

We once told a Year 7 class that we would tell them how to save half a million dollars each. They were somewhat disappointed when all we said was, 'Don't smoke!' However, once we worked out together the lifetime financial cost (not to mention health cost), they all seemed to have made a sound decision based on understanding of rate: $20 per packet × 2 packets per day × 365 days per year × 40 years = $580 000.

Economists talk about 'opportunity cost' (the benefits we forego by choosing to spend money in a particular way). Imagine the opportunities that could be taken advantage of if this kind of money was available during your life! Understanding rate is an important element of number sense that helps us to make decisions in our lives. Often, as we've seen in the examples in this section, it is the impact of the rate over periods of time that leads to positive or negative outcomes.

## Learning activity 8.11

Building on the previous examples, think about some of the regular expenses you have in your life; some of course are necessities but perhaps some are discretional. Now, get a calculator and work out the long-term financial impacts of some of these. In particular, consider small but regular rates of expenditure (e.g. phone plans, pay TV subscriptions, takeaway meals instead of homemade lunches) that occur over a period of time. They are the ones that can 'fly under the radar' if we are not consciously using our number sense.

The final section of this chapter focuses on another important application of relative and multiplicative thinking, which is commonly used in a range of situations in our daily lives.

# SCALE

Understanding scale is a highly underrated skill, and we probably use scale in our lives a lot more than we realise. The term 'scale' has a number of meanings. For example, the term can:

- refer to the relative size of objects
- refer to the axes on a graph or marked intervals on a measuring instrument
- provide information about a map (e.g. scale factor)
- refer to one-, two- and three-dimensional objects.

Note that the term scale in this chapter does not refer to scales that are used to measure mass.

A detailed look at some of the applications of scale listed above is beyond the scope of this book, and rather than try to provide a single definition we will use a number of examples. In order to work successfully with scale in the contexts referred to in this chapter, we will apply ideas we have already explored in this book, such as number facts, multiplicative thinking, relative thinking, fractional thinking, and ratio. Again, using these ideas to think about situations involving scale will depend on and contribute to our number sense.

## Relative size

In Chapter 6 we discussed relative thinking. Sometimes scale is useful to help us think relatively to imagine or understand magnitude. For example, the size of an object or the depth of water on a road crossing is often indicated through the use of a scale, as shown in Figure 8.6.

**Figure 8.6** Examples of scales for indicating magnitude or depth

In the photo on the left in Figure 8.6, the dolphin has been made to represent the actual size of a live dolphin. The scale has been added so that people can more clearly understand it. Looking at the photo, can you imagine how tall you would be next to the dolphin? The photo on the right in Figure 8.6 shows a roadside scale indicating the depth of flood water. Why would this be more useful than a sign that simply indicates that the road is subject to flooding?

## Reading scales

The ability to read and interpret linear scales (e.g. rulers, thermometers, graph axes) is essential to using numerous measuring instruments, as well as more formal skills such as reading and interpreting graphs. We will start with a familiar example: the scales on the side of a measuring jugs, such as those shown in Figure 8.7.

**Figure 8.7**   Two different kitchen measuring jugs

Both jugs in Figure 8.7 have intervals marked in steps of 50 mL. Note that the jug on the right does not label every 50 mL, although they are marked, whereas the jug on the left has labels for every 50 mL. Neither jug shows intervals for every millilitre as this would be impractical. Using increments of 50 mL allows us to make reasonable judgements of the amount of liquid needed. If very accurate measures to the nearest millilitre are needed, we would rely on our numeracy skills relating to selection and use of appropriate mathematical tools to choose a different measuring container; for example, we might use an eye dropper or medicine glass for medication or measuring spoons in the kitchen.

What is important when using scales successfully is to understand what each interval means. The intervals were clearly marked on the jugs shown in Figure 8.7, but this is not always the case and it can take time and application of multiplicative and relative thinking to become proficient at reading and understanding scale. Children get experience of this when working with intervals on number lines. Being given the first and last number, they are often asked to fill in some missing values. Consider the number line in Figure 8.8, which requires an understanding of scale, and the information it provides.

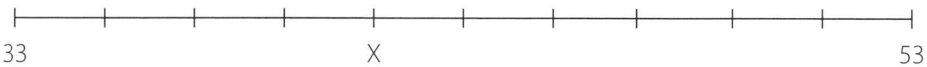

33                                          X                                          53

**Figure 8.8**   A number line segment

We used the number line segment shown in Figure 8.8 in a study of over 2 500 middle-year students and asked them to determine the value of X. To correctly answer this question, relative thinking is required because the two end values (33 and 53) need to be considered along with the number of intervals between them. The relative thinkers were able to correctly decide that there are 10 intervals representing a difference of 20 (i.e. 53 − 33) so each interval must have a value of 2. This tells us that the value of X = 41 (25% of students from Year 5 to 9 used relative thinking to answer this correctly). The absolute thinkers didn't take account of the 53 and instead counted in ones from 33 to obtain an incorrect answer of X = 37 (36% did this).

# Learning activity 8.12

1.  On the number line below, give the value of X.

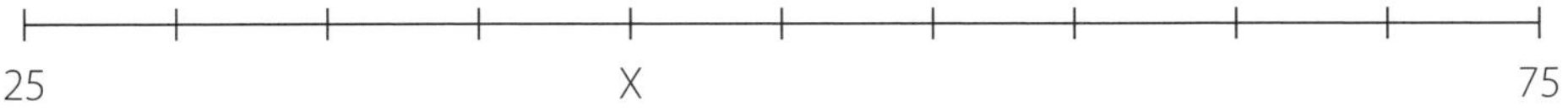

25                                          X                                          75

2.  What measurement is indicated on the scales on the objects in these photos for: (a) the width of the light switch? (b) the time remaining in minutes? (c) the size of the angle?

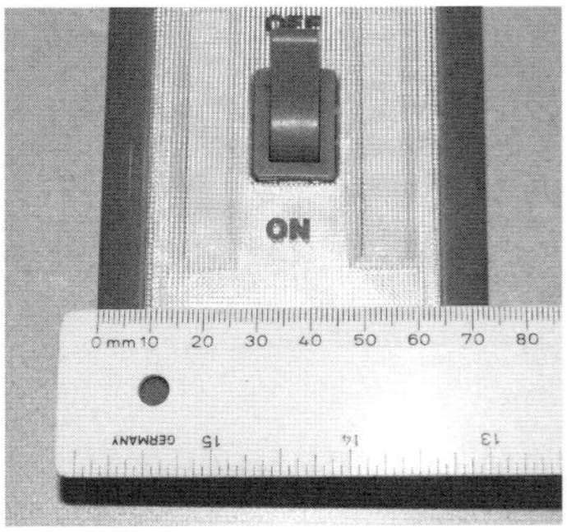

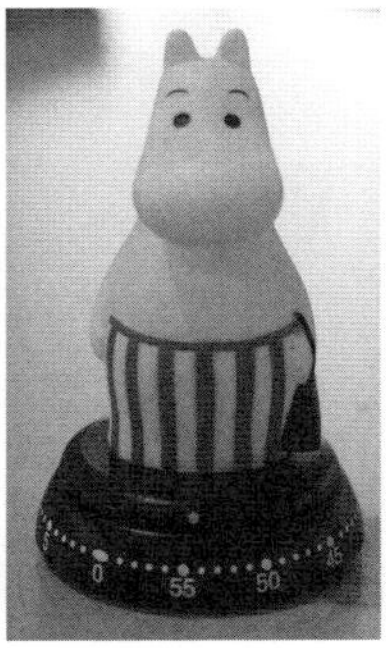

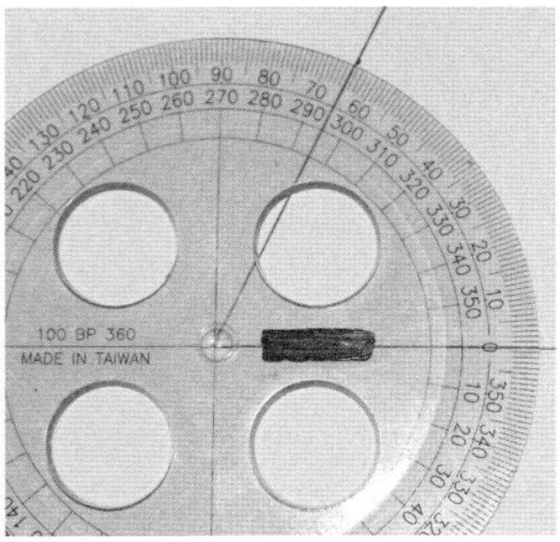

*(Answers in Appendix.)*

While some of the above examples might seem like standard mathematics questions, there are many applications of scale reading in everyday life (some of which we have already mentioned). We need the skills used in the previous examples for such activities as interpreting the axes on graphs, interpreting timelines and reading measuring instruments (from the kitchen to the car; working on a building site to working in a lab). Stop and think about when and where you might use these skills in your own life. Scenario 8.7 relates to the everyday life of a teacher.

Scenario 8.7

## TEACHING CREATIVE WRITING

A teacher we worked with recently was researching a six-week writing intervention with her Year 3 class. The intervention aimed to teach the students some skills for improving aspects of their creative writing. She gave the students a writing assessment before she started her intervention (the pre-intervention assessment) and again after she had completed the six weeks of teaching (the post-intervention assessment). Her results are shown in Figure 8.9.

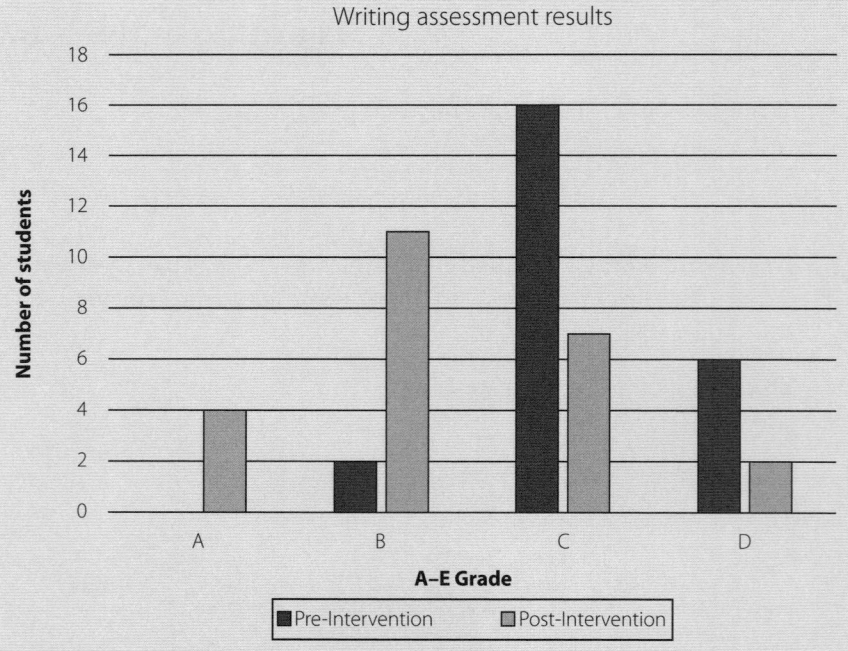

**Figure 8.9**　Pre- and post-intervention results for Year 3 writing

In Learning activity 8.13 you used relative thinking to interpret the vertical axis and then you used additive thinking (subtraction in this case) to find the answer to the question. Reading scales and interpreting or producing graphs are probably the two most commonly used applications of scale requiring our number sense in everyday or work situations. There are two other applications of scale that we will now look at briefly because they are related to number sense and to everyday applications of the concepts we have explored so far in this book: scale factor and one-, two- and three-dimensional scale.

## Scale factor

Scale factor is used to indicate how many times larger or smaller a diagram or model is than the actual entity it represents. It is common to see scale factors on maps, house plans and scale models in museums. Scale factors are usually represented as ratios. Some may include the units for ease of interpretation: for example, a regional tourist map in a brochure might have a scale of 4 cm : 20 km. Other scale factors do not include units: for example, an architect's house plan might simply say 1:1 000 and we know that every 1 mm on the plan represents 1 m (i.e. 1 000 mm) in reality. Map reading is not as common as it once was now that we have access to digital maps on our phones, computers and satellite navigation in our cars. The maps on these devices (and on some hard-copy maps too) tend not to use ratio representations of scale. Rather, they have a linear scale with increments (e.g. centimetres) that indicate real distance, as shown in Figure 8.10. How would this be used?

0               125               250               375 m

**Figure 8.10**   Example of scale factor on maps

## Learning activity 8.14

1.  It is interesting to see what happens to the scale representation on digital maps. On your phone, open a maps app. Using your fingers, zoom in and zoom out on the map. As you do this, a scale factor diagram similar to that in Figure 8.10 should appear. Watch what happens to it as you zoom in or out on the map.
2.  The distance between towns A and B on my paper map is 3 cm. The scale tells me that 1 cm equates to 15 km. If the road between the towns is relatively straight, approximately how far apart are they in reality? *(Answer in Appendix.)*

It is useful to use scale factor when planning things around the house because drawings to scale allow you to make practical decisions. For example, you might draw a room plan for relocating or purchasing new furniture to make sure it will fit; you might be renovating the bathroom and want to estimate the amount of tiles or paint you will need; or you might be designing a garden or landscape layout.

## Learning activity 8.15

In this activity you will practise creating a scale drawing of a room. Imagine that you are planning to rearrange the furniture in a room where you live (e.g. a bedroom or living room). Draw a simple scale drawing of the room using a scale of 1:20 (this means that 1 cm on your plan will represent 20 cm in reality). You can choose whether to include the furniture in its current location or to recreate your room.

## One-, two- and three-dimensional scale

Some of the ideas around this topic are not so commonly used in everyday situations, but they are used by certain professions, such as, architects, engineers and builders, and many aspects are beyond the scope of this book because we are focusing on commonly encountered situations that require your number sense. In the following examples we briefly explore some of the ideas in this topic.

### One-dimensional scale

We have already spent time interpreting one-dimensional scales when we read number lines, read measuring instruments and interpreted the axes on graphs. One-dimensional scales are sometimes referred to as linear scales. An example commonly used in school is a timeline in history, which is used to show sequences and relative times of events. Sometimes history timelines are not drawn to scale because their purpose is simply to show the order of a series of events. This can have an impact on students' understanding of the time periods involved so it is preferable from a numeracy point of view to ensure that they are drawn to scale whenever possible.

## Learning activity 8.16

Draw a timeline from 1900 to 2000 to scale to indicate the timing of the following events of the 20th century. Before you start, decide on the scale you will use and what an appropriate interval might be.

- World War 1          1914–1918
- World War 2          1939–1945
- Sputnik launch       1957
- Moon landing         1969
- Fall of Berlin Wall  1989
- End of Cold War       1991
- Your birth (you may have to extend the timeline)

*(Answer in Appendix.)*

## Two-dimensional scale

This aspect of scale refers to objects that are two-dimensional, such as maps, photographs, plans, and plane shapes in geometry. Two-dimensional scale is also useful when we are only interested in two dimensions of an object. An example is deciding what I will need to do to the recipe if I want to use a cake pan that is twice as long (but with the same depth) as the one in my recipe. Scenario 8.8 has an example similar to the cake pan idea. Note that although cake pans and tiles (in Scenario 8.8) are three-dimensional objects, we only need to consider two of their dimensions in these examples).

## Scenario 8.8

## TILING THE BATHROOM

I decided to tile an area in the bathroom, which measures 4 m × 4 m. I had the choice of square tiles that were either 40 cm × 40 cm or 20 cm × 20 cm. I decided use 20 cm × 20 cm tiles because they were half the price of the 40 × 40 cm tiles. Did I get a good deal? My neighbour said I didn't!

$\rightarrow$

While it might seem that each 20 cm × 20 cm tile is half the size of each 40 cm × 40 cm tile, we need to consider not just the length of one side but both dimensions. Because each side of the 40 cm × 40 cm tile is double the side of the 20 cm × 20 cm tile, it is the area of the tile that I should have thought about. Look at the photo.

It turns out that each 40 cm × 40 cm tile will cover four times the area of the 20 cm × 20 cm tile. So, did I get a good deal or not? (No because the large tile was four times the small tile but only two times its price.)

## Three-dimensional scale

When working with three-dimensional scale, we consider the dimensions of three-dimensional objects. Many students struggle with this aspect of scale. Think about this situation: I make small gift boxes to hold homemade chocolates as presents for friends. If I want to make a box to hold twice as many chocolates, what will I need to do to the box's dimensions? In Figure 8.11 we've used blocks to simulate this situation.

**Figure 8.11**   Blocks to illustrate impact of changing dimensions of a box

On the left of Figure 8.11 you can see that we have one block. If we double one dimension, we will double the capacity of the box, as shown by the second image. What is the impact of doubling two dimensions? In the final image, each of the dimensions has been doubled (i.e. 2 blocks long by 2 blocks wide by 2 blocks high). By doubling all three dimensions, we have changed the number of blocks by a factor of 8. So, if I doubled all three dimensions of my chocolate box, I would need to put eight times the number of chocolates in the box.

## Not to scale

Sometimes models, diagrams or maps may not be to scale, and this is usually implied or indicated explicitly. For example, 'mud maps' or hand-drawn maps drawn quickly to give someone directions may focus on where to turn left or right but not the actual distances between. What number sense would someone need to use to interpret such a 'mud map'? We have a diving map of Julian Rocks, a local diving site. The map explicitly states that the map is 'not to scale'. Why would this be important information for divers?

Sometimes scales can change. For example, in Figure 8.12 you can see the speedometers of our two cars. One has the intervals labelled consistently every 20 km/h. The other starts with the intervals labelled every 10 km/h and then changes to 20 km/h. What are the implications for someone who didn't realise the scale had changed?

**Figure 8.12**  Two different car speedometers. Note the different use of intervals on each.

There are situations in which we may be asked to use a notional scale to answer a question. For example, a doctor might ask us what level of pain we are in on a scale from 1 to 10. A teacher might ask a child how serious is the problem that they are reporting;

for example, 1 = I can solve it myself, 5 = I can solve it with my friends, 10 = I need an adult to help. These are scales that are subjective, but their relative nature helps clarify the severity of the situation. The patient or the child needs number sense to respond effectively.

# CONCLUSION

In this chapter we have focused on three important areas of mathematics: ratio, rate and scale. They all rely on numerous elements of number sense, including many from previous chapters. In particular, a good understanding of ratio and rate relies on multiplicative and relative thinking as well as an understanding of fractions. Scale is multifaceted and, in many situations, it relies heavily on relative thinking. While we have provided some examples to show the diversity of its applications, we have tried to restrict the ideas to those that might be relevant to most people's everyday lives. As always, we conclude with our suggestions to improving your own number sense in relation to ratio, rate and scale.

## Personal actions to improve number sense

By following Polya's problem-solving strategy, it is possible to now *carry out the plan* to improve your personal number sense. The following are suggestions to assist you:

- Identify moments in your life where a ratio needs to be adhered to.

- Understand the order in which a ratio is expressed, and how it might also be expressed as a fraction or percentage.

- Interpret graphic depictions of situations using ratio.

- Identify aspects of your life that could be measured as a rate or that involve rates.

- Identify rates that you feel are positive for you and any rates that you feel would be beneficial to change.

- Identify your use of scale as one-, two- and three-dimensional.

- Notice any scales in everyday situations or contexts.

# 9

# Problem solving

## LEARNING OBJECTIVES

After reading this chapter, you should be able to:

- recall and understand what is involved in the four steps of Polya's problem-solving process
- apply Polya's four steps to a range of problems
- understand the importance of mathematical language and multiple representations to problem solving
- understand the importance of number sense as a contributor to successful problem solving.

# INTRODUCTION

This final chapter deals with problem solving. Here we will draw on the ideas from the previous chapters to look at the nature of problems and to consider ways to solve them. Problem solving is one of the four proficiency strands in the Australian Curriculum and is described as the ability to interpret, formulate, model and investigate by applying a variety of strategies in unfamiliar situations, and verifying that answers are reasonable (ACARA, 2020a).

This description does not mention number sense explicitly and yet number sense is central to being an effective problem solver. Our goal is to help readers to continue to develop their own number sense. A key reason for needing and using number sense is to solve mathematical problems – the focus of this chapter.

There are many ideas about what constitutes a problem in mathematics and even more on what makes a 'good problem'. We will stick to simple ideas. A problem can be considered anything that challenges someone to solve it without predetermined or rehearsed strategies. There may be more than one way to obtain the answer and what is most important is the application of mathematical thinking. This is one of the reasons that we have left this topic until now. In this chapter we will hone in on the steps that you can take to tackle a range of problems. While there are many ideas about how best to achieve this goal, we will use Polya's problem-solving steps because you are already acquainted with them from our discussion throughout this book.

# PROBLEM SOLVING AND LIFE

Throughout this text various skills and strategies have been explored to help improve your personal number sense and in turn your personal numeracy. Remember that numeracy is the use of mathematics in context. It is in problem solving that the impact of improved number sense will be most felt. There is an old saying that 'you don't really know something till you teach it', similarly in mathematics we like to think that 'you don't really know something until you apply it'! Problem solving in real life in many cases is the everyday application of our number sense.

You probably remember 'problem solving' from school. It usually entailed some form of written word problem that you had to interpret and then figure out what mathematical knowledge to apply to the situation. Remember this: What is the quotient when the sum of 1 380 and 368 is divided by the difference between 906 and 883? Of course, any gaps in mathematical skills (e.g. number facts or specific terminology) or strategies (e.g. relative thinking) meant that success would be limited. The result of this for many students was that the spiral of maths anxiety or disengagement was exacerbated. Often the problems to solve seemed irrelevant and disengaging: 15 books each 1.2 cm thick and 28 books each 1.5 cm thick just fill a shelf. Find the length of the shelf.

Schulz, the famous Charlie Brown (*Peanuts*) cartoonist, summed it up best through his character Sally Brown: 'Only in Math problems can you buy 60 cantaloupes, and no one asks what is wrong with you.' In contrast to some of the problem examples considered in previous chapters, real life is full of 'problems', which are relevant and can't really be avoided, and many involve mathematics. To solve life's problems, one must be as confident and competent as possible with applying mathematical knowledge to the specific context.

Problem solving in the real world is largely characterised as knowledge creation. Such problems can be personal (e.g. reorganising a weekly budget) while others can be internationally earth-shattering (e.g. finding a cure for a disease). When solving personal problems we may not be creating knowledge that is new to others, but it is new knowledge to the problem solver, thus an important personal achievement.

## Reflection activity 9.1

Describe any problems or situations in the past week that have required you to use your mathematical skills and knowledge (e.g. budgeting, planning a holiday, shopping, paying bills).

Most if not all financial decisions will be unique to your situation and will likely involve some level of problem solving related to money – for example, budgets, interest, time, rate, loans and credit card payments. There are many different mathematical variables to consider, calculate or balance. Similarly, many health situations involve problem solving or the application of mathematical ideas – for example, medication dosage rates, exercise times and rates, and adjustment of diet.

## Problem solving and growth mindset

Problem solving requires strong understanding of the basics underpinning a situation and positive dispositions and a growth mindset. These attributes develop over time as success breeds success. Think about driving. Initially, you learned the skills and began applying them in real life after you obtained your licence. After a few years you generally became a more skilled driver with greater traffic sense. Having gained the initial driving skills, most people become better at problem solving in traffic, perhaps most importantly being able to anticipate or avoid dangerous situations. This progression of becoming a problem-solving driver is not automatic: we all know some people who remain 'scary' drivers all their lives. One doesn't have to spend too long on the road to realise that there are two types of drivers: problem solvers and problem creators! Relating this scenario to personal problem solving using our number sense, just as with driving we need to attain the basic skills and then persist with their application to all the varied situations that life brings us.

One of the important skills that problem solvers have is that of reflection. Thinking about the success or lack thereof in solving a problem helps prepare us for the next, just as reflection on mistakes or on other experiences can help us to learn more deeply. Because problems are so varied, it is not possible to have a 'one size fits all' set of solutions. Like an expert driver, as problem solvers we need to arm ourselves with skills and attitudes and then reflect on the outcomes.

This chapter will look at some typical mathematical 'problems' but we will also consider some real-life 'problems' to solve. The term 'problem' in a 'life' context generally has negative connotations, so while we will continue to use the word in this chapter, in life it might refer to an issue, a challenge or even some positive circumstance. However, because life is full of 'problems' it is generally considered that one's success at/in life (whatever that means to you) entails some skills at solving problems. Not all of life's problems entail mathematics, but it seems a surprising number do. Think of any situations where you have to consider time, measurement, money, ratios, fractions, calculations, scale and so on. Some of life's situations might be 'mathematics free', such as emotions or relationships, though sometimes even these may have mathematical undercurrents – for example, partners sharing finances or time or workload.

In mathematics a problem can be thought of as a task or activity that students are asked to undertake and for which they have no perception that there is a 'correct' method for finding the solution or for which they have not memorised a set of rules (Van De Walle, Karp & Bay-Williams, 2014). This description of what constitutes a mathematical problem suggests that if there are no prescribed or memorised rules for solving a problem, then a set of other mathematical skills are needed to assist us to solve mathematical problems.

# CONTRIBUTING SKILLS FOR PROBLEM SOLVING

Before we start working through the steps in Polya's problem-solving process, we will describe a number of important skills and abilities that a successful problem solver uses.

## The language of problem solving

When children engage in 'problem solving' in a classroom mathematics lesson, one of the more persistent difficulties they have is with the language used. We have had some teachers say to us that they see this as a literacy problem not a numeracy one. We tend to disagree with this. Mathematics is not just numbers, but it also has associated language. Just as with the examples of percentages in Chapter 7, the language of mathematics is quite nuanced and specific. Equal emphasis must be given to the numbers and the words. As stated earlier, the problem with problems is that no two are alike so each scenario in mathematics and life is unique. Before solving a problem with numbers, the language must be interpreted and yet just as problems are diverse, so too is the language used and it is often context-specific. In other words, having a strong understanding of mathematical terms and language is important to successful problem solving.

To illustrate why the diversity of terms and language used can be a challenge, here are some examples of varied language for the four main operations:

- *Add:* sum, more, as well as
- *Subtract:* difference, less, minus, take away
- *Multiply:* product, times, lots of
- *Divide:* quotient, shared equally, split equally among.

Of course, there are myriad terms in mathematics that are used for solving certain problems beyond the four operations.

## Learning activity 9.1

Here are some words or phrases from textbook problems. Some are quite specific to a particular operation, whereas others may be applicable to more than one operation, depending on the context of the problem. For each of the following terms or phrases, decide which of the four operations they refer to.

| | | |
|---|---|---|
| • tally | • bisect | • fewer |
| • how long | • deduct | • doubled |
| • how many altogether | • decrease by | • twice as much as |
| • groups of | • cut equally | • partition |
| • quarter | • withdraw | • halve |
| • by how much | • how much left | • how much did each |
| • trebled | • total | • how many more than |

*(Answer in Appendix.)*

So, it is important to remember that the language of problems is often the more difficult element to interpret and we should always seek clarity, especially with the 'fine print'.

## Multiple representations

We have already considered many representations in this book. In the previous section we focused on textual or linguistic representations of mathematical ideas. In Chapter 7 we discussed the ability to represent the same number as a common fraction, a decimal fraction or a percentage. In that chapter we also represented fractions on number lines, using collections of counters, and through diagrams and photos. Some representations of one-quarter are shown in Figure 9.1. **Multiple representations** are defined as different ways of representing aspects of the same idea or concept (Hilton & Nichols, 2011). In simple terms, they are exactly what they sound like: lots of different ways of representing something.

> **Multiple representations** are different representations of aspects of the same concept; for example, they may be verbal, graphical, numerical or diagrammatic.

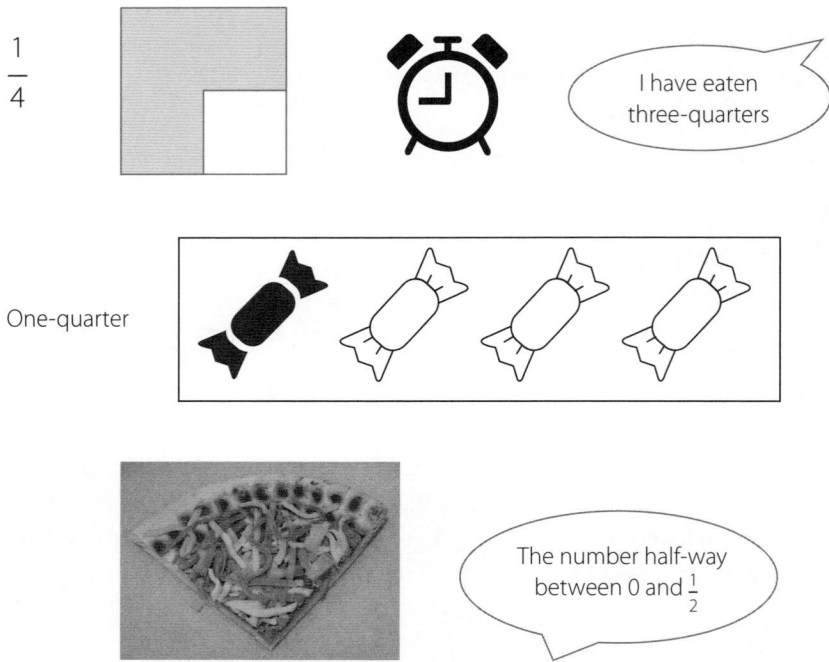

**Figure 9.1**    Multiple representations of one-quarter

## Learning activity 9.2

1.  What different information is portrayed by each of the representations in Figure 9.1?
2.  Create two of your own representations of one-quarter that differ from those in the figure.

It is important to be able to use, interpret, analyse and create multiple representations because sometimes multiple representations are needed to fully represent a concept. This is especially important if a concept is quite abstract because a single representation may not adequately portray all the required information.

In the context of problem solving, multiple representations can be used to visualise a problem or to represent or interpret information. If the problem is especially complex or involves multiple steps, using representations can help us to understand the problem in a different way. Some categories of multiple representations are listed below. Think of more examples of when you use some of these types.

- Textual/linguistic representation: written words that represent mathematical ideas
- Symbolic: the 'language of mathematics' used to represent mathematical entities or relationships (e.g. +, ÷, $2 \times 3$, $x^3$, km/h, {2, 4, 6 ...}, =)
- Diagrams: drawing a plan to scale with measurements
- Tabular: making data tables (e.g. at tax time for various expenses or deductions)
- Pictorial: photos to visualise a situation
- Gestural: moving around (e.g. stepping out the lounge to imagine new furnishings)
- Physical: using manipulatives, such as blocks or counters to represent a situation
- Verbal: discussing with a friend to clarify things in your mind
- Graphical: creating and interpreting graphs (e.g. to track spending on phone bills or electricity bills)

The trick with multiple representations is to select the one/s that best suit the problem at hand.

## Learning activity 9.3

Look again at the previous list of representational types.

1. Identify a time when you have used one or more of the listed representational types to interpret or solve a mathematical situation or problem.
2. Describe how you used the representation(s) to solve the problem.

## Thinking mathematically

Another important skill set when solving problems is thinking mathematically. This includes being able to explore, question, work systematically, visualise, make conjectures, explain, represent, generalise, justify or prove. The NRICH website (Cambridge University,

2020) provides clear examples of these ideas. According to Cambridge University (2020), to develop mathematical habits of mind we need to be resourceful, collaborative, resilient and curious. Of course, being able to think mathematically also involves number sense – something that we have emphasised throughout this book.

If the basics such as understanding of the Hindu-Arabic number system and knowledge of number facts have improved and awareness of possible strategies (mental computation, additive and multiplicative thinking, absolute and relative thinking, fractional thinking, and ratio, rate and scale) have been explored, then approaching problem solving can be done with improved confidence. It still does not give a silver bullet for success, but the important thing is an attitudinal expectation that one is better prepared and more likely to find success.

In Chapter 1 we introduced Polya's steps to problem solving as a framework referred to throughout this text to help improve personal number sense. In a piece of circular logic, we will again work through these steps in the context of problem solving. A reminder that the steps are: (1) Understand the problem; (2) Devise a plan; (3) Carry out the plan; (4) Look back at what you've done. It can be seen that these are quite broad, mainly because no two problems are the same so specificity is difficult. In the following sections we will build on the ideas presented so far to work through Polya's problem-solving process.

## POLYA'S STEP 1: UNDERSTAND THE PROBLEM

Thinking back to our school days it is difficult to remember any time in mathematics classes that we were asked to identify a problem. The basic approach to 'problem solving' was to give students a predetermined problem and then tell them to work it out. However, as we have seen, the first step to solving a problem is to understand or identify it, which sounds simple but often it is not a clear-cut situation. Effectively, many of us have gone through school without ever experiencing the first step of problem solving.

Understanding a problem in life sometimes requires some uncomfortable honesty. Trying to solve financial problems, for instance, will require some very clear understanding of where money is being spent and what our personal priorities are. While mathematics may not be so personal, being able to devise a plan to solve any problem, mathematical or not, requires full and clear understanding of the problem. The very first step is identifying what the question is or what the problem requires you to find out.

For example, the problem might be finding the most cost-effective phone plan for my circumstances. Understanding this problem requires us to know that the solution of the problem will result in a cost-effective phone plan. To understand and find a solution to the problem, we need to identify (and find) relevant facts. For example, we may need to know facts such as how much credit a plan provides for local and overseas calls, how much roaming data is provided, and the overall monthly cost. We would need to know these facts for multiple providers to make an informed choice. We also need to understand which of the features of the plan are actually relevant to our personal circumstances (e.g. will we be making calls overseas?).

Understanding a problem requires us to know that we have all the relevant facts or information. However, sometimes 'facts' aren't always clear and obvious. Sometimes we have more information than we need – too many facts. At other times there may be information missing that we will need to find by using our mathematical thinking and representation skills. In some cases we can even have the 'wrong' or conflicting facts.

In the phone plan example, there are particular **parameters** (characteristics of an individual's needs regarding a plan and their personal use of their phone). Deciding on the parameters can help to identify which facts are relevant and this is especially the case when problems are open-ended or ill-defined (as is the case with many real-world problems) or when we have many facts to consider. For example, the phone plan user may only use their phone on Wi-Fi networks, so they won't want to pay for roaming data they won't use. They may have a strict budget and want to set an upper price limit on their choice.

> **Parameters** are limits or boundaries that define a problem, process or activity.

Sometimes it is also helpful to identify any **assumptions** that are needed to help define or clarify the problem and this can again help us to decide which facts are relevant. Again, in the context of the phone plan example, an incorrect assumption might be that the chosen network would cover you in the areas in which you would commonly use your phone.

> **Assumptions** are used to clarify or define a problem so that the solution is meaningful.

In the following sections we look at how the varying levels of access to facts can affect our problem solving.

## Irrelevant facts (red herrings)

When solving problems we are presented with an array of information, some of which may be pertinent and some may not; being able to tell the difference is an ever increasingly important critical numeracy skill in these days of fake news and false claims. The irrelevant information found in a problem-solving situation is sometimes referred to as a 'red herring' as it has the potential to mislead the problem solver. The important issue is that one must be aware that in life problems and mathematics problems the possibility exists that not all information available will help solve the problem.

Returning to the first step in problem solving, understand the problem, clearly identifying exactly what the problem is asking you to answer can be helpful in sifting out the irrelevant facts (as in Question 1 in Learning activity 9.4).

## Too much information

This is another case where it is essential to keep your focus on what you are actually required to find out. Let's return to the phone plan example. If you collected all the information supplied by numerous providers (e.g. 6 providers), each with multiple phone plan options (e.g. 6 plans), you could be forgiven for feeling that you had too much information and were becoming overwhelmed by the decision among 36 possible plans! This is also an example of a problem where it would be helpful to decide on your parameters so that you can eliminate some plans quickly.

## Missing information

Problems are very hard to solve if we don't have all the information. In a way this situation is the opposite of the 'red herring' scenario. Often problems can be so complex that it is difficult to clearly see all the information or to even realise that some is missing. As mentioned earlier, knowing all the facts in the fine print of a financial document is vital; it is not wise to make decisions if we have not accessed all the information.

At a local council election there were three candidates: A, B and C. A received 500 of the total votes; B received 350 more votes than A. How many votes did C receive?

2.  I make an international money transfer. I look for the best deal. I decide that I will use the company that has 'no commissions'. What other information would I need to know before being confident I had the best deal?

Hopefully in the second question of Learning activity 9.5 you identified the need to know more information about the exchange rate offered by the money transfer company compared to that of others. It may be that the offered rate has been reduced to compensate for the zero commission charges. Other information might include the time taken for the transfer or the security offered by the company.

In a mathematical situation it may be that sometimes it is not possible to have all the information at the beginning of a problem. For example, we have a colleague who asks her students, 'What makes a great paper plane?' or 'How far can an origami frog jump?' These are very open problems for which many approaches and strategies might be useful. While wondering about paper planes and origami frogs is not an everyday occurrence, the point is that often real-world problems are quite open and not clearly defined. The first step in understanding such problems is defining them and this is where assumptions and parameters may be useful. They can help us to refine and define the problem so that we can understand the problem in more detail and have a clearer idea about the information we will need to solve it. For example, in the case of the origami frog we might assume that the frog is made from a particular size of paper; we might state that we will measure the 'jumps' from a line to its nose and that the direction of the jump is not important, just the distance. The paper plane problem can be refined by deciding what we mean by a 'great' paper plane; for example, is it the distance it can fly or the height it can reach or some other parameter? These are examples of situations where we need to provide ourselves with more information about the problem.

## Not understanding all the information

As soon as we step outside our sphere of understanding in our world, we can sometimes feel illiterate or innumerate in that situation. We recently listened to a podcast about a trainee pilot; the language used to describe many of the technical issues was completely unknown to us, a situation where we were functionally illiterate. When deciding on a new phone plan, our eyes glazed over, mildly overwhelmed by the complexity of choices, even those involving numbers: type of phone, memory size/cost, contract with/without own phone, length of contract, rates per call, rates per minute, varying costs for different phone companies, discounts, free call limits, overseas call costs and so on. Making decisions about phone plans, for us at least, is always somewhat frustrating as we are unclear about some of the meanings of the pertinent information, especially when different companies sometimes use different terminology to refer to similar things. We try our best but, in the end, we usually give up and go with the same company we always did and trust the sales assistant to give us an honest and fair deal. Trying to understand all the facts of a situation is vital to successful problem solving. Luckily these days we can more readily seek and find information we need by going online.

Once the problem is clearly identified and is understood the next step is devising a plan for solving the problem.

# POLYA'S STEP 2: DEVISE A PLAN

Once we are satisfied, to the best of our ability, that all the relevant facts of a situation are known and we understand what the problem is asking, a plan can be devised. As we have seen, Polya's framework is very broad and provides only a general guide to problem solving. As problems are quite often unique, with very particular facts, it is impossible to give a one-size-fits-all means of devising a plan. Instead, it is important to have a range of strategies that you can use when and if they are relevant and it is at this point that our number sense can help. Sometimes we need a few other strategies to help us apply our number sense. Ultimately, the strategy will depend on the nature of the problem and it is fair to say that whatever you choose to do, you will need to decide on the mathematical steps necessary to solve the problem. In short, you will need to transform your representations of the problem into mathematical symbols so that you can carry out your plan. In the following sections we will look at some useful strategies and then consider some scenarios to illustrate them.

## Using multiple representations

Most problem-solving situations begin with consideration of whether one or more representations might be helpful. For example, would it be useful to

- draw a diagram or picture of the situation?
- use physical materials to help model the problem?
- act out the situation (either individually or with others)?
- make a table, graph or chart?
- make a list of all the information first?
- write an equation or number sentence?

Being able to visualise a problem can be a helpful strategy when working out a problem. It can help you to understand what the problem is about and sometimes it can help you visualise a strategy or solution or at least decide what a reasonable strategy or solution might be. This step can help you to decide what to do next (e.g. What operations will you use? Will you need more than one step to solve the problem? What assumptions should you make?). Sometimes, one or more representations will help you to clarify a problem and this will lead to deciding what to do next or what operations will be useful.

# Learning activity 9.6

You may remember your teachers asking you to do this type of problem. It is a very old textbook problem (you'll tell from the prices).

A couple had $500 to buy new furnishings. They spent $120 on a lounge suite at the first store. They then bought a dining suite for $310 at the next store. Can they afford the bedroom suite for $200 at the next shop? Can you think of a way of representing this situation to help you visualise what is happening? Draw a quick picture or other representation (you might like to try writing an equation or number sentence too).

*(Answer in Appendix.)*

If a teacher used the problem in Learning activity 9.6, they might have asked the students to *visualise* what was happening to the amount of money, perhaps by asking some children to act out the situation (i.e. gestural representation). By doing this, children would hopefully have a better understanding of the key words in the problem (i.e. spent, bought, afford). This would also help them decide what operations were required to complete the problem. While as adults we probably won't act out problem scenarios, the visualising of the situation to help develop a clear understanding is something we all do.

## Deciding what operation(s) to use

As discussed earlier, the language used in problems, both in the mathematics classroom and in real life, can be quite varied and nuanced and this can make it difficult to decide what operation or operations to use to move towards an answer. For instance, the exchange rate conversions in previous chapters present this challenge – do I multiply or divide? Understanding the information in the problem and also what the problem is asking of you can help.

Sometimes no obvious operation is articulated. Think about this problem from an old textbook: How many envelopes each containing $46 can be filled with $414? To work out the answer to this strange (and contrived) question, we might need to imagine or act out the action of filling the envelopes; the word 'filled' in the question is acting as an instruction to divide $414 by $46.

Before we continue looking at strategies, we will look at Scenario 9.1 to consider the use of visual representations to help understand the problem and choose operations.

## Scenario 9.1

### THE WINE RACK

I have 2 identical wine racks, each of which has 3 rows with 4 spaces in each row. Most of the spaces contain a wine bottle but I notice that there are 5 empty spaces altogether. I don't want to count all the bottles that I have, but I need to know how many bottles there are. How can I do this?

This scenario has many words to unpack and while most are not mathematical in nature, it is easy to see how someone who is not feeling confident might feel overwhelmed. Think how much easier the question would be with a picture or diagram, as shown in Figure 9.2.

**Figure 9.2**  Image of wine racks missing 5 bottles

Did seeing the pictorial representation help you to understand the question or visualise the situation? Try to link the words in the problem to the information in the photo. What do you think the next step might be?

In Scenario 9.1, seeing the picture helped us to understand what we needed to find out. This allows us to select operations to solve the problem. Notice that the wine racks look like the arrays we discussed in Chapter 4 when we looked at multiplication, so this tells us what we need to do. First, we need to know how many wine bottles the wine racks can hold when full by multiplying $8 \times 3 = 24$ (or $2 \times 4 \times 3$). We then find out how many are left if 5 bottles are gone (subtraction $24 - 5 = 19$). Which step was additive and which was multiplicative?

## Working backwards

A good way of determining whether a mathematical concept is understood is to be able to work backwards from an answer. Teachers often employ this strategy with their classes. This can also be a useful strategy in solving some problems – both in mathematics and in life. How often do you receive a bill (the answer) and you try to work backwards through it to find out where the money was spent (think restaurant bill, phone bill or credit card bill)? The following example requires the solver to 'undo' the problem. Try to solve this before reading on. *I think of a number, double it and then add 8 to the result. The answer is 46. What was the original number?* This problem requires us to reverse the operations and the order. Starting at 46, subtract 8 (38) then halve 38 (19). If you enjoyed this type of 'Think of a number' problem, you can find any number to practise online.

---

### Learning activity 9.8

Here is another problem to illustrate the idea of working backwards: You buy 5 identical lengths of timber for a garden project. The total cost, including $35 delivery, is $110.

1. What is the cost of the timber?
2. What about the cost of each piece of timber?
3. What part of this required additive thinking and what part required multiplicative thinking?

*(Answers in Appendix.)*

---

So far we have looked at using representations, choosing operations and working backwards. Sometimes problems can be more challenging because they have multiple steps in them.

# Multistep problems

Problems that have multiple steps to a solution can be notoriously difficult. This is because a number of decisions about operations (+, −, ×, ÷) or other mathematical procedures have to be made as well as about the order of completion. An error in any part will lead to an overall error. Of course, real-life situations can have many steps to a solution and require lots of planning and thought for a successful outcome. Many of our students identify multistep problems from school days as a catalyst for negative mathematical feelings. Managing multiple facts with multiple operations in a variety of possible sequences makes it very easy for an error to occur. Lack of success can leave students disappointed and feeling unwilling to further engage. However, with improved number sense one could expect that past experiences could be slowly overcome. A good strategy for complex problems can be to make a list of steps and follow them to solve the problem. Scenario 9.2 illustrates some strategies to solve multistep problems.

## Scenario 9.2

### PAINTING MY LIVING ROOM WALL

I want to repaint the end wall of my living room. It is shaped like a *trapezium*. The *length* at the floor is *5 m* and the *heights* of either end are *3 m and 5 m*. My paint has directions that tell me I can paint *16 m² with 1 L of paint* (i.e. a *rate* of *16 m² per litre*). The current colour is quite dark so I will need *at least two* coats. I have a *4 L* can of paint. Will this be enough paint?

The problem in Scenario 9.2 contains a lot of information. We have italicised all the words that require knowledge of mathematics or number sense. When we first read the problem, we may identify the need to understand a number of different things in order to solve the problem, including geometry, measurement, rate and numerical operations. The problem is also a multistep problem. Let's look at some strategies that might help us.

The first thing we could do is to *make a representation* – probably a diagram. This would help us understand the shape and size of the wall (as shown on the left in Figure 9.3). Most of us don't remember much about trapezia (plural of trapezium), so we might choose to break the diagram down further to make a rectangle and a triangle, as shown on the right in Figure 9.3.

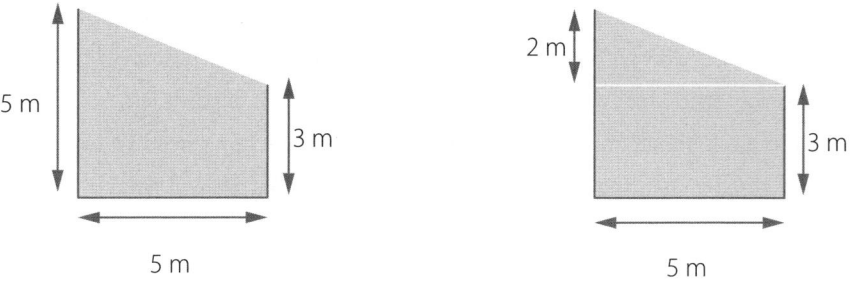

**Figure 9.3** Two representations of the wall in Scenario 9.2

It may not be immediately obvious from the diagram what we need to do next. We can think about the problem as a series of steps and *make a list*. This can be a useful strategy in a multistep problem because we can think about the operations in each step. The list for Scenario 9.2 might look like this:

- Calculate how much area we will have to paint (area of rectangle + area of triangle).
- Calculate how much area we will paint in two coats (area from previous step × 2).
- Find out how many litres we will need (1 L of paint covers 16 m² so I will divide the answer to the previous step by 16 to find out how many litres I need).
- Compare the previous answer to the size of our paint can to decide whether we have enough. (Is my answer to the previous step more or less than 4 L?)

While the list we just made may have left some readers feeling confused or frustrated, stay with us because there are more strategies we can try. Let's rewind to the idea of *working backwards*. Scenario 9.2 is a real-life situation that we encountered recently. We didn't actually sit down and draw diagrams and calculate areas of rectangles and triangles. This is where number sense and a focus on what the problem is actually asking you to find out are important. Starting at the end point, we know we have 4 L of paint. From here we simply want to know whether that is enough paint to do the job, not the exact number of litres needed. Look at the second diagram from Scenario 9.2 again:

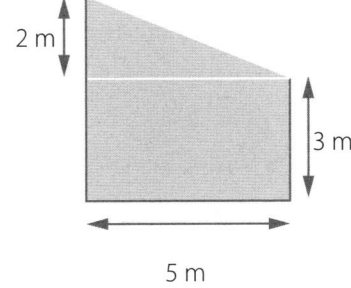

**Figure 9.4** The second diagram

We know we have 4 L of paint. The rectangle's area is 15 m² (3 × 5). This is less than the 16 m² that we can paint with 1 L. That means that we will need less than 2 L to paint the rectangle twice. After that, we will have a little more than 2 L of paint left.

The triangle is much smaller than the rectangle, so we don't need to find its area – because it's smaller than the rectangle, we already know that we will need even less than 1 L of paint for each coat of that section.

Answer to the problem: We will have more than enough paint for two coats.

In the second approach to the problem in Scenario 9.2 we used number sense rather than strict mathematical calculations. If we had been asked to calculate exactly the number of litres needed, we would have needed to use the series of steps. Also note that in both approaches, we made the assumption that we would only apply two coats. How would the situation differ if we needed three coats?

## Other strategies

There are some other strategies that might be useful in solving problems and some of these relate to ideas we already discussed. For example, you might find it useful to highlight mathematical language (as we did in Scenario 9.2) before deciding what operations or strategies are needed. You might want to try looking at a similar but simpler form of the problem and determining how that could be solved and then going back to the original question. That might have worked with the paint question by using a small regular shape and simpler numbers than 16 m². You can also use guess and check (trial and error approaches), which sometimes require you to work backwards to decide whether your solution makes sense.

## Unfamiliar contexts

Immersing ourselves in an unfamiliar context can be difficult. There is often a little nervousness or apprehension associated with the new situation. This can occur the first time we do something, such as trying downhill skiing, arriving in a new country or applying for a bank loan. Sometimes this unfamiliarity can block our clear thinking.

A student of ours had trouble with a science problem. It was hard to understand why because the problem was not overly difficult, and the student was a high achiever. When asked why he was having trouble, he replied, 'I don't know what geomorphology is'. The unknown context had hindered his usual acute thinking. The lesson here is that even though we may not be familiar with the context, we can often still apply our skills if we can overcome our initial discomfort.

# POLYA'S STEP 3: CARRY OUT THE PLAN

Carrying out a plan to solve a problem is of course crucial and it clearly relies heavily on the plan you devised in Step 2. We can think of numerous times in our lives when problems have been identified and plans made, but when it came to the crunch, no action was taken. This of course involves the 'hope for the best' strategy, or the 'wait long enough and the problem will go away' technique, neither of which are really recipes for success. Here are some ideas that may help in carrying out the plan to solve a problem.

## Thinking about possible outcomes

It is always helpful if we have a notion of what the outcome of a plan might be; both in mathematics and in life more generally. Using skills of estimation can help with this. Estimation gives us a guide as to what to expect or what might be feasible. It also gives us a focus to keep working on enacting our plan. It can also alert us to situations in which our plan might be going off track.

We do have to ensure that these thoughts of a possible outcome don't distort our thinking; in a way becoming a self-fulfilling prophecy. Just desiring a particular outcome is not enough to bring it to fruition.

If you're tackling a multistep problem, it can be helpful to keep checking the list you originally made. Are you making the progress you thought you would? Are your answers so far what you predicted? This approach can help you to avoid a lot of unnecessary work because it can sometimes indicate where a plan needs to be modified.

# POLYA'S STEP 4: LOOK BACK AT WHAT YOU'VE DONE

When a solution to a problem has been determined, it is very important to review the processes and answer. Reasonableness of answer is important, especially if estimation skills (see Chapter 5) were employed at the beginning. The answer then becomes a confirmation of what you were expecting. Everyone is fallible, so it is reasonable to expect that we will quite often make mistakes, but there are some things in life where we really must be certain of a correct outcome. So, to do this we need to confirm our calculations by:

- re-identifying the problem
- checking the facts again
- checking strategies and calculations
- checking for reasonableness of answer.

A classic but sad example of not completing this final review stage was a joint space program (the 1999 Mars Climate Orbiter) being conducted between the United States and the United Kingdom. Too late, vital calculations were found to be incorrect because of confusion between one country's use of the imperial measurement system and the other's use of the metric system. This led to the loss of the $125 million spacecraft.

An important aspect of many real-world problems is that they can be 'open-ended', meaning that a number of solutions may be possible, and different solutions may be more suitable to some people than others. Think about the decision to rent or buy a house. Different people, with different circumstances, will make different decisions. At different times of life different choices may be suitable. Think about some similar decisions that you and your contemporaries have to make where different choices or outcomes are available; did you choose the same internet plan, phone plan or type of car?

# AN AUTHENTIC EXAMPLE OF PROBLEM SOLVING

In this final section we present an example that draws together the elements of problem solving that require number sense and numeracy to solve the problem. The aim of this section is to illustrate how they are interconnected and can be drawn upon.

## Scenario 9.3

### MAKING A PATHWAY

*Understand the problem:* A narrow section of our yard was shady and often muddy. To address this, we decided to build a pathway. The problem therefore was determining what materials to use and what layout to use.

*Devise a plan:*

- We used string to mark out the edges of the pathway (physical representation).
- We walked the length of the proposed pathway to determine the number of paces (additive strategy, gestural representation, estimation).
- We converted the number of paces to the required number of stepping stones – number of stepping stones = number of paces + 1 for the beginning of the path (additive thinking, mental computation).
- We decided on the dimensions of the stones using the width of the path and the number of stones needed (number sense – relative size, understanding relationships).

*Carry out the plan:* We placed the first and last stones and then placed the remaining stones so that they were equidistant, adjusting where necessary, before we embedded them using sand and gravel.

*Look back at the results:* The completed pathway looked the way we wanted, allowed easy walking because of the size and spacing of the stones. Most importantly, it solved the problem of the muddy area.

A similar set of steps was required to determine how much bedding sand was required under the stepping stones, the amount of gravel needed, and the length of the garden edging.

## CONCLUSION

This chapter focused on problem solving as an important aspect of number sense. Because of the nature and diversity of problems, both in life (everyday, work, civic and so on) as well as in more formal mathematical situations, it is not easy to present a simple approach to solving them. Instead, we have described the four steps of Polya's problem-solving process as a way of helping readers to approach problems in a systematic way and to show that positive dispositions and strong number sense can help to achieve our goal of solving them.

As a final word, we would like to congratulate the readers of this book for staying the course with us and for having the determination and positive dispositions to want to improve their number sense. We know that the journey is not always easy and that some of the ideas in this book may still need more focus or practice. We encourage you to continue to work on building your number sense and remember that growth mindset!

## Personal actions to improve number sense

By following Polya's problem-solving strategy, it is possible to now *carry out the plan* to improve your personal number sense. The following are suggestions to assist you:

- Be conscious of the steps you take to solve everyday problems.

- Identify the elements of number sense you use to solve problems.

- Be aware of other aspects of problem solving (e.g. using representations, knowing all the facts, multistep problems, working backwards and so on).

- Look at the reasonableness of your solutions to problems.

And finally, the fourth step of Polya's process involves looking back. We encourage you to look back (and reflect) on the progress you've made while engaging with the ideas in this book to improve your personal number sense.

# APPENDIX

## Answers to learning activities

### CHAPTER 2

#### Learning activity 2.2

1. $7 \times 15 = 105$ chocolates (perhaps you thought $7 \times 10 + 7 \times 5$)
2. $35 - 19 = 26$ (perhaps you thought $35 - 20 = 25$, so subtracting 19 instead gives 26)

In situation 1 you used multiplication and in situation 2 you used subtraction.

#### Learning activity 2.3

1. Using estimation, the area is approximately $7 \times 3 = 21$ m². I have 2 L of paint, which would allow me to cover 32 m². Assuming I only want to paint the wall once, I have enough but if I need to paint more than one coat, I will need to buy more paint.
2. Rockhampton is about 800 km away and Townsville is just over 1 500 km so I will stay overnight in Rockhampton. (This is a little over half-way.)

#### Learning activity 2.4

1. a. 400
   b. 0.14
   c. 23.34
2. $65 \div 4$ does equal 16.25 but we can't have 0.25 of a person, so we must round down to 16 people.
3. This calculation is logical in that I am running 100 times as far (100 m $\times$ 100 = 10 000 m or 10 km) but if we are being realistic, I won't be able to run at the same pace over 10 km that I could over 100 m so it will take longer than 1 300 seconds to run 10 km.

## Learning activity 2.5

```
      3   4.  5
 +  2 3   4.  1   5
          9.  1   2   3
 ──────────────────────
    2 7 7.  7   7   3
 ──────────────────────
```

Place value allows us to use setting out in which we arrange the numbers so that we can add accurately.

## Learning activity 2.6

I want to make three times as many muffins, so I need to multiply the amount of each ingredient by 3. This means I will need to multiply 1.5 cups by 3 (so I will need 4.5 cups of flour).

## Learning activity 2.8

1.  I know that 20% is the same as one-fifth. One-fifth of $150 is $30. If I will save $30, I will need to pay $120. (If you didn't know that 20% = $\frac{1}{5}$, another way to think about the question might be that 10% of $150 = $15 so 20% is $30.)

2.  To compare the costs, I need to calculate how much it will cost me per day.

| $\frac{1}{2}$ day | $69 | $138 per day |
|---|---|---|
| 2 days | $120 | $60 per day |
| 3 days | $165 | $55 per day |
| 4 days | $200 | $50 per day |
| 5 days | $240 | $48 per day |

The cheapest option per day is the five-day lesson package. Using my numeracy skills, I would also want to decide whether I need (and can afford) to take lessons for five days.

# CHAPTER 3

## Learning activity 3.2

1.  32 509 354: thirty-two million, five hundred and nine thousand, three hundred and fifty-four

2. 1 254 098 465: one billion, two hundred and fifty-four million, ninety-eight thousand, four hundred and sixty-five

3. 908 196 740 031: nine hundred and eight billion, one hundred and ninety-six million, seven hundred and forty thousand, and thirty-one

## Learning activity 3.3

1. 34.5: thirty-four and five-tenths
2. 487.23: four hundred and eighty-seven and twenty-three hundredths
3. 1 305.004: one thousand, three hundred and five, and four thousandths

# CHAPTER 4

## Learning activity 4.1

b. $7 + 8 = 15$; $8 + 7 = 15$; $15 - 8 = 7$; $15 - 7 = 8$

c. $11 - 8 = 3$; $11 - 3 = 8$; $8 + 3 = 11$; $3 + 8 = 11$

d. $17 - 9 = 8$; $17 - 8 = 9$; $8 + 9 = 17$; $9 + 8 = 17$

## Learning activity 4.3

Each giraffe has four legs. Skip counting: 4, 8, 12; Repeated addition: $4 + 4 + 4 = 12$; Multiplication facts: $3 \times 4 = 12$

## Learning activity 4.5

1. A and C (the other two statements are incorrect; e.g. $10 - 2 \neq 2 - 10$)
2. C (the additive identity, i.e. 0, does not change the value of the number when added)
3. B (the multiplicative identity, i.e. 1, does not change the value of the number when multiplied)

## Learning activity 4.9

a. $8 \times 20 = 160$
b. $720 \div 9 = 80$
c. $720 \div 90 = 8$
d. $8 \times 0.3 = 2.4$

**e.** $18 \div 0.6 = 30$

**f.** 30% of \$600 = \$180

**g.** $\dfrac{1}{4} \times 36 = 9$

## Learning activity 4.10

205 m + 125 m

$$
\begin{array}{r}
2\ \ 0\ \ 5 \\
+\ \ 1\ \ 2^1\ 5 \\
\hline
3\ \ 3\ \ 0 \\
\hline
\end{array}
$$

The distance between the two sites is 330 m.

(Another approach might be to think of 205 as 200 + 5 and add 200 + 125 = 325 and then add the remaining 5 to get 325 + 5 = 330 m)

# CHAPTER 5

## Learning activity 5.3 (all use the Distributive Law)

**1.** $4 \times 8 + 6 \times 8 = 10 \times 8 = 80$

**2.** $3 \times 7 + 4 \times 3 = 11 \times 3 = 33$

**3.** $8 \times 6 - 6 \times 6 = 2 \times 6 = 12$

**4.** $\$25 \times 3 + \$25 \times 7 = \$25 \times 10 = \$250$

**5.** $6 \times 15 = 6 \times 10 + 6 \times 5 = 60 + 30 = 90$

## Learning activity 5.4

Here are some of our thoughts on how the calculations in Question 4 of this learning activity may have been done in your head. If the methods you used were in any way different, then consider whether they were efficient and effective.

**a.** $26 + 34 = 60$

Here we're excited because the ones column is a number fact adding to 10:

- $(4 + 6) + (20 + 30)$ Same order as pen and paper
- $(20 + 30) + (4 + 6)$ Tens column first
- $(26 + 4) + 30$ Maintain the first number, add the ones column followed by the tens column

**b.** $47 + 38 = 85$

This is more difficult because we have to retain a carry figure from 7 + 8:

- (7 + 8) + (40 + 30) Same order as pen and paper
- (40 + 30) + (7 + 8) Tens column first
- (47 + 30) + 8 Maintain the first number, add the tens column followed by the ones column

c.  65 − 24 = 41

Here again we're excited because each digit of 24 is less than corresponding in 65:

- (5 − 4) + (60 − 20) Same order as pen and paper
- (60 − 20) + (5 − 4) Tens column first
- (65 − 4) − 20 Subtract ones from total then tens

d.  71 − 28 = 43

This is more difficult because 8 is not less than 1:

- (71 − 8) − 20 Subtract ones first then tens
- (71 − 20) − 8 Subtract tens first then ones
- (71 − 21) − 7 Subtract 21 because it works well with 71, then subtract the 7 that is left
- (71 − 30) + 2 Subtract an 'easy' round number like 30, then compensate by adding back the 2

e.  23 × 5 = 115

We're not too worried because it hovers around familiar 25 × 5 or 20 × 5:

- (3 × 5) + (20 × 5) Same order as pen and paper*
- (20 × 5) + (3 × 5) Tens column first*
- (25 × 5) − (2 × 5) Round up to 25 because it is easy to multiply and then subtract to extra 2 × 5*

    (*Notice that all of these ideas use the Distributive Law: 3 lots of 5 + 20 lots of 5; 20 lots of 5 + 3 lots of 5; 25 lots of 5 − 2 lots of 5)

f.  43 × 8 = 344

There aren't so many strategies for this; just remember that 40 × 8 is a number fact (10 times 4 × 8):

- (3 × 8) + (40 × 8) Same order as pen and paper
- (40 × 8) + (3 × 8) Tens column first

    (Again, these strategies use the Distributive Law.)

g.  147 ÷ 7 = 21

This is exciting because you can recognise some numbers that work well together:

- $(140 \div 7) + (7 \div 7)$ Same as pen and paper $(14 \div 7$ number fact)

**h.** $87 \div 3 = 29$

A bit tricky because we have to keep a remainder in our heads and think place value:

- $(8 \div 3)$ is 2 remainder 2 (but remember it is tens column, so 2 goes in the tens column for the answer and remainder 20 goes with the 7 to make 27) $27 \div 3 = 9$. So, $20 + 9 = 29$. Okay, a bit long winded but basically do it in your head the same as pen and paper.
- An approach that should remind you of the Distributive Law:

    $87 \div 3 = (60 + 27) \div 3$. Knowing your number facts and your knowledge of multiples let you calculate this as:

    $60 \div 3 + 27 \div 3 = 20 + 9$

    $= 29$

**i.** $97 + 112 + 3 = 212$

In this question we hope you used associativity because you recognised that regrouping first would simplify the question to $97 + 3 + 112 = 100 + 112$ and then using your place value knowledge, the answer is 212.

**j.** $4 \times 20 \times 15 = 1\,200$

While the first part of this calculation is easy as written $(4 \times 20 = 80)$, the second step $(80 \times 15)$ may not be so straightforward. This is another example where application of associativity and place value can help. Regrouping allows us to change the calculation to:

$4 \times 15 \times 20 = 60 \times 20$ and then we use our knowledge of place value:

$60 \times 20 = 1\,200$

(In this last step remember that we can think of $60 \times 20$ as $6 \times 2 \times 10 \times 10 = 12 \times 100$, which is $1\,200$.)

## Learning activity 5.7

**a.** $1\,000$ Danish Krone $(1\text{AUD} = 4.8$ DK$)$

This calculation can be rounded to $1\,000 \div 5 \approx 200\text{AUD}$ ($\approx$ means 'approximately equal to').

**b.** 36 USD $(1\text{AUD} = 0.72$ USD$)$

This calculation relies on your recognising a number fact $(35 \div 7 = 5)$.

Rounding 36 USD to 35 USD and rounding the exchange rate to 0.7 simplifies our calculation to $35 \div 0.7 = 50$ AUD (remember place value and the fact that when we divide by a number smaller than 1 our answer is larger).

c. 50 South African Rand (1 AUD = 12.3 SA Rand)

There are two probable approaches to this calculation:

**(1)** $50 \div 12.3$ is approximately equal to $48 \div 12 = 4$ AUD

**(2)** $50 \div 12.3$ is approximately $50 \div 10 = 5$ AUD

The estimate here depends on your choice of rounding strategies. The actual answer is 4.06 AUD.

# CHAPTER 6

## Learning activity 6.4

| Program | Joining fee (paid once) | Monthly fees | Total cost for six months |
|---------|-------------------------|--------------|---------------------------|
| Bronze | $20 | $10 | $80 |
| Silver | $30 | $20 | $150 |
| Gold | $40 | $30 | $220 |

Calculating the total monthly fees for 6 months was multiplicative; including the joining fee was additive.

## Learning activity 6.5

Absolute amounts: 25 m long; 100 tonnes; 100 000 kg

Relative thinking: comparison of the mass of the tongue to that of an elephant; comparison of the size of the heart to the size of a car.

## Learning activity 6.6

Examples might be:

1. If I increase the distance I run, it will take more time.
2. If I increase my speed, I will get there in less time.
3. Decreasing the amount of screen time per day will increase my time for other activities.
4. Decreasing my shower time will decrease my water consumption.

## Learning activity 6.7

1. If the usual dosage is 2 tablets then for a double strength medication, the dosage should be 1 tablet.
2. The situation is not relative. Dosages are determined after a great deal of testing and they are stipulated with good reason. It is very dangerous to vary recommended dosages so this is an absolute situation.

## Learning activity 6.8

1. a. The matchbox gives the viewer something to compare to the fish. It makes it easier to understand the length of the fish.
   b. Because the matchbox is being used as a point of comparison with the fish, the fisher is invoking relative thinking.
   c. The fish is approximately 8 times longer than the matchbox so it is approximately 48 cm (you may have said a little more than this because we can't see the end of the tail so any answer between 48 and 55 cm would likely be reasonable).
2. a. The height of the statues by themselves is difficult to determine because there is no point of reference.
   b. When we know the man is 2 m tall, the nutcracker statue is about the same height as the man (2 m). The museum statue is about twice the height of the man (4 m).

## Learning activity 6.9

1. $109 + 425 + 1\,690 = 2\,224$
2. Men

| People | Children | Women | Men |
|---|---|---|---|
| On board | 109 | 425 | 1 690 |
| Saved | 56 | 316 | 338 |
| 3. Perished (absolute number) | $109 - 56 = 53$ | 109 | 1 352 |
| 4. Percentage perished (relative to those on board) | $53 \div 109 \times 100 = 48.6\%$ | 25.6% | 80% |

5. a. In absolute terms there were more men saved than the other groups.
   b. In relative terms there were more women saved than the other groups (this is because the percentage of women that perished is lower than for the others).

c. In absolute terms more men perished than the other groups.

d. In relative terms more men perished than the other groups.

It does appear to be a case of women and children first because the number of men who died is highest no matter whether we look at it absolutely or relative to the total number on board.

## Learning activity 6.10

1.

| Passenger class | Passengers on board | Passengers saved | Passengers lost |
|---|---|---|---|
| Second | 285 | 118 | 285 − 118 = 167 |
| Third | 706 | 178 | 528 |

2. Percentage of second-class passengers saved = 118 ÷ 285 × 100

   = 41.4%

   Percentage of third-class passengers saved = 178 ÷ 706 × 100

   = 25.2%

3. a. In absolute terms there were more third-class passengers saved.

   b. In relative terms there were more second-class passengers saved.

   c. In absolute terms more third-class passengers perished.

   d. In relative terms more third-class passengers perished.

# CHAPTER 7

## Learning activity 7.1

1.

2.

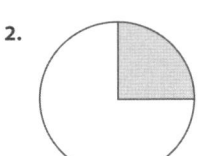

## Learning activity 7.2

**1.**

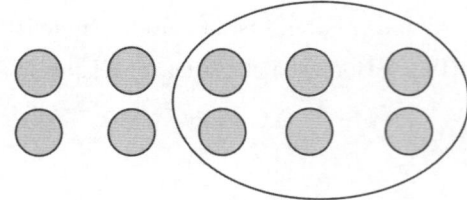

**2.**

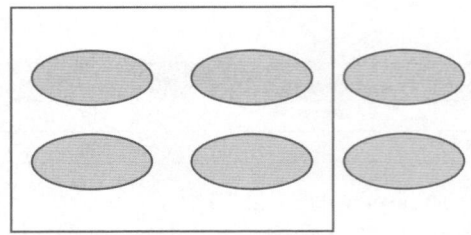

## Learning activity 7.3

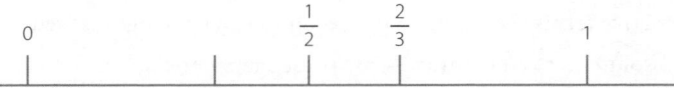

## Learning activity 7.4

**1.** $5 \div 9 = \dfrac{5}{9}$

**2.** $23 \div 37 = \dfrac{23}{37}$

## Learning activity 7.5

Patrick would say that it's not fair because he is now getting $\dfrac{5}{6}$ of Charlie's pocket money per hour. Put another way, for every $5 that Patrick earns per hour, Charlie earns $6 per hour.

## Learning activity 7.6

**1. a.** Lowest common multiple of 5 and 10 is 10

**b.** Lowest common multiple of 4 and 6 is 12

2. Equivalent fractions:

   **a.** $\dfrac{1}{2}$ and $\dfrac{1}{6}$ are equivalent to $\dfrac{3}{6}$ and $\dfrac{1}{6}$

   **b.** $\dfrac{1}{3}$ and $\dfrac{2}{5}$ are equivalent to $\dfrac{5}{15}$ and $\dfrac{6}{15}$

## Learning activity 7.7

1. **a.** $\dfrac{5}{16}$      **b.** $\dfrac{1}{4}$
2. **a.** 5      **b.** 16
3. **a.** 18      **b.** 20

## Learning activity 7.8

1. 516
2. 2.3
3. 4
4. 0.7

## Learning activity 7.9

1. 0.08, 0.18, 0.8
2. 0.01, 0.04, 0.1

## Learning activity 7.10

1. 0.5
2. 0.3
3. 0.55
4. 0.035

## Learning activity 7.11

1. $\dfrac{7}{8}$ and $\dfrac{16}{18}$ as percentages: 87.5% and 88.9% so, $\dfrac{16}{18}$ is larger

2. $\dfrac{107}{300}$ and $\dfrac{46}{139}$ as percentages: 35.7% and 33.1% so, $\dfrac{107}{300}$ is larger

## Learning activity 7.12

1. $3.50
2. $18
3. $42
4. 14.35
5. $2.25
6. 320 km
7. $180

## Learning activity 7.13

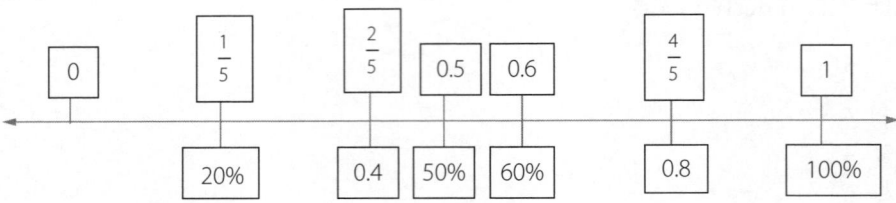

## Learning activity 7.14

1. The survival rate was presented as a percentage because this is easier to relate to than 163 of 837.
2. It makes it easier to visualise and compare the number of survivors and the number who died.
3. No, because it is easier to think about 81% than an unfamiliar fraction.

# CHAPTER 8

## Learning activity 8.1

1. **a.** Lemon trees : orange trees = 100:400. Lowest terms 1:4
   **b.** Fraction of trees that are orange trees = $\dfrac{400}{500}$ or $\dfrac{4}{5}$
2. **a.** cats : dogs = 30:25 or 6:5 in lowest terms
   **b.** fraction of dogs = $\dfrac{25}{55}$ or $\dfrac{5}{11}$ in lowest terms

# Learning activity 8.2

1. 15
2. 20
3. 10:25 or 2:5 in lowest terms

# Learning activity 8.3

The message is that a lot more people in the community need blood than the number who donate blood. The aim of the signage is to encourage more people to be blood donors.

# Learning activity 8.4

In general the more nurses/midwives in a population, the lower the infant mortality rate.

# Learning activity 8.5

1. The 9 L can needs 30 mL of fertiliser.
2. The 1.5 L can needs 5 mL of fertiliser.
3. The 2 L can needs 6.7 mL.

# Learning activity 8.7

1. 75 km/h
2. 3 hours

# Learning activity 8.8

25 accidents per year on average.

# Learning activity 8.9

The varying rates are used because small things are sold in punnets rather than by kilogram (e.g. blueberries); some large items (e.g. rockmelons) are sold as individual items. The nature of the fruit and vegetables (e.g. size, packaging) determines how it is priced for sale.

## Learning activity 8.10

1. Car A 24 L of petrol; Car B 15 L of petrol
2. Car A $36; Car B $22.50
3. Car B saves $13.50
4. Car A uses 3 L per km more fuel than Car B, which at $1.50 per litre makes a difference of $4.50 per hundred kilometres. Multiply 100 km by 3 000 to get 300 000 km for the life of the car. Now multiply the $4.50 by 3 000 to find the overall fuel cost difference. The owner of Car B saves $13 500 over the life of the car!

## Learning activity 8.12

1. X = 45
2. a. 73 mm
   b. between 54 and 55 minutes (note that this timer reads in the opposite direction to what we would usually expect because it counts down to zero)
   c. 65 degrees

## Learning activity 8.13

1. 11 students
2. 16 – 7 = 9 students

## Learning activity 8.14

2. 45 km

## Learning activity 8.16

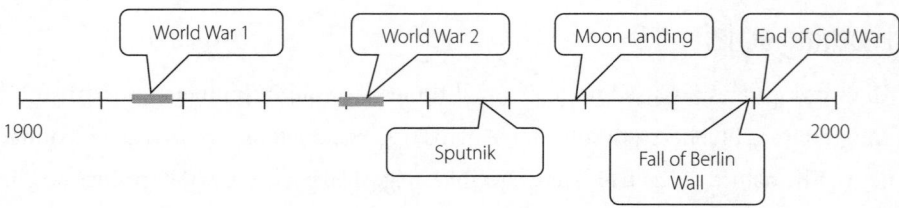

# CHAPTER 9

## Learning activity 9.1

*Addition:* tally, total, how long, how many altogether

*Subtraction:* deduct, decrease by, withdraw, how much left, by how much, fewer, how many more than

*Multiplication:* doubled, trebled, groups of, twice as much as

*Division:* partition, cut equally, halve, quarter, bisect, how much did each

Depending on the context, some of these words might be used with more than one operation.

## Learning activity 9.4

1. The car weighs 2 tonnes is the red herring (it is irrelevant).

## Learning activity 9.5

1. You need to know the total number of votes.

## Learning activity 9.6

The answer is no, they can't buy the bedroom suite.

## Learning activity 9.7

(Sale price − deposit) ÷ 12 = monthly repayment

## Learning activity 9.8

1. Cost of timber = $110 − $35 = $75
2. Cost of each piece of timber = $75 ÷ 5 = $15
3. Additive thinking (subtraction) to find the cost of the timber; multiplicative thinking (division) to find the cost of each piece.

# REFERENCES

ACARA. (2020a). Mathematics proficiencies. Retrieved from https://www.australiancurriculum .edu.au/resources/mathematics-proficiencies/

ACARA. (2020b). National Numeracy Learning Progression. Retrieved from https://www .australiancurriculum.edu.au/resources/national-literacy-and-numeracy-learning-progressions/national-numeracy-learning-progression/number-sense-and-algebra/?subE lementId=50723&searchNodeId=50725&searchTerm=climate+change

ACARA. (n.d.). Numeracy. Retrieved from https://www.australiancurriculum.edu.au/f-10-curriculum/general-capabilities/numeracy/

Adelson, J. L. & McCoach, B. (2011). Development and psychometric properties of the Math and Me Survey: measuring third through sixth graders' attitudes toward mathematics. *Measurement and Evaluation in Counselling and Development*, 44(5), 225–47.

Attard, C. (2011). 'My favourite subject is maths. For some reason no-one really agrees with me': student perspectives of mathematics teaching and learning in the upper primary classroom. *Mathematics Education Research Journal*, 23(3), 363–77.

Bandura, A. (2001). Social cognitive theory: an agentic perspective. *Annual Review of Psychology*, 52(1), 1–26.

Barkatsas, A., Kasimatis, K. & Gialamas, V. (2009). Learning secondary mathematics with technology: exploring the complex interrelationship between students' attitudes, engagement, gender, and achievement. *Computers and Education*, 52, 562–70.

Beilock, S. & Willingham, D. (2014). Maths anxiety: can teachers help students reduce it? *American Educator*, Summer, 28–43.

Berch, D. B. (2005). Making sense of number sense: implications for children with mathematical disabilities. *Journal of Learning Disabilities*, 38(4), 333–9.

Blackwell, L. S., Trzesniewski, K. H. & Dweck, C. S. (2007). Implicit theories of intelligence predict achievement scores across an adolescent transition: a longitudinal student and an intervention. *Child Development*, 78(1), 246–63.

Booker, G., Bond, D., Sparrow, L. & Swan, P. (2014). *Teaching primary mathematics*. Frenchs Forest: Pearson.

Brady, K. & Winn, T. (2017). *Maths skills for success at university*. Melbourne: Oxford University Press.

Bright, G. W., Joyner, J. M. & Wallis, C. (2003). Assessing proportional thinking. *Mathematics Teaching in the Middle School*, 9(3), 166–72.

Brown, J. S., Collins, A. & Duguid, P. (1989). Situated cognition and the culture of learning. *Educational Researcher*, 18(1), 32–42.

Burnett, S. & Wichman, A. (1997). *Mathematics and literature: an approach to success*. Chicago, IL: Saint Xavier University and IRI/Skylight. Retrieved from https://files.eric .ed.gov/fulltext/ED414567.pdf

Cambridge University. (2020). Thinking mathematically. Retrieved from https://nrich.maths .org/mathematically

Claxton, G. (2014). *School as an epistemic apprenticeship: the case of building learning power.* The 32nd Vernon-Wall Lecture presented at the Annual Meeting of the Education Section of the British Psychological Society. Retrieved from https://www.education.sa.gov.au/sites/default/files/school_as_an_epistemic_apprenticeship.pdf?acsf_files_redirect

Council of Australian Governments. (2008). *National numeracy review report.* Retrieved from https://web.archive.org.au/awa/20080719174118mp_/http://www.coag.gov.au/docs/national_numeracy_review.pdf

De Klerk, J. (2014). *Pearson illustrated maths dictionary* (5th edn). Melbourne: Pearson Australia.

Dewey, J. (1931–32). American education past and future. In J. A. Boydston (ed.). *The later works* (Vol. 6, pp. 90–8). Carbondale, IL: Southern Illinois University Press.

Dole, S., Clarke, D., Wright, T. & Hilton, G. (2012). Students' proportional reasoning in mathematics and science. In T. Tso (ed.), *Proceedings of the 36th Conference of the International Group for the Psychology of Mathematics Education* (Vol. 2, pp. 195–202). Taipei, Taiwan: PME.

Dowker, A., Bennett, K. & Smith, L. (2012). Attitudes to mathematics in primary school children. *Child Development Research,* https://dx.doi.org/10.1155/2012/124939

Downton, A., Russo, J. & Hopkins, S. (2019). The case of disappearing and reappearing zeros: a disconnection between procedural knowledge and conceptual understanding. In S. B. G. Hine & A. Cooke (eds.), *Mathematics education research: impacting practice* (Proceedings of the 42nd Annual Conference of the Mathematics Education Research Group of Australasia) (pp. 236–43). Perth: MERGA.

Eccles, J. S. & Roeser, R. W. (2011). Schools as developmental contexts during adolescence. *Journal of Research on Adolescence*, 21(1), 225–41.

Geiger, V., Forgasz, H. & Goos, M. (2015). A critical orientation to numeracy across the curriculum. *ZDM Mathematics Education*, 47, 611–24.

Goos, M. (2007). Developing numeracy in the learning areas (middle years). Keynote address delivered at the South Australian Literacy and Numeracy Expo, Adelaide.

Goos, M., Dole, S. & Geiger, V. (2011). Improving numeracy education in rural schools: a professional development approach. *Mathematics Education Research Journal*, 23, 129–48.

Goos, M., Vale, C., Stillman, G., Makar, K., Herbert, S. & Geiger, V. (2017). *Teaching secondary school mathematics: research and practice for the 21st century.* Crows Nest, NSW: Allen and Unwin.

Grant, H. & Dweck, C. S. (2003). Clarifying achievement goals and their impact. *Journal of Personality and Social Psychology*, 85(3), 541–53.

Gray, J. J. (2019). Mathematics. Retrieved from https://www.britannica.com/science/mathematics

Haylock, D. (2019). *Mathematics explained for primary teachers.* London: Sage Publications.

Hilton, A. (2018). Engaging primary school students in mathematics: can iPads make a difference? *International Journal of Science and Mathematics Education*, 16, 145–65.

Hilton, A., Hilton, G., Dole, S. & Goos, M. (2013). Development and application of a two-tier diagnostic instrument to assess middle years students' proportional reasoning. *Mathematics Education Research Journal*, 25, 523–45.

Hilton, A., Hilton, G., Dole, S. & Goos, M. (2016). Promoting students' proportional reasoning skills through an ongoing professional development programme for teachers. *Educational Studies in Mathematics*, 92, 193–219.

Hilton, A., Hilton, G., Dole, S., Goos., M. & O'Brien, M. (2012). Evaluating middle years students' proportional reasoning. In J. Dindyal, L. Chen & S. Ng. (eds). *Mathematics education: expanding horizons* (pp. 330–7). Singapore: MERGA.

Hilton, A. & Nichols, K. (2011). Representational classroom practices that contribute to students' conceptual and representational understanding of chemical bonding. *International Journal of Science Education*, 33(16), 2215–46.

Hogan, J. (2002). Mathematics and numeracy – is there a difference? *Australian Mathematics Teacher*, 58(4), 14–16.

Hughes-Hallett, D. (2001). Achieving numeracy: the challenge of implementation. In L. A. Steen (ed.). *Mathematics and democracy: The case for quantitative literacy* (pp. 93–8). USA: National Council for Education and the Disciplines.

Lamon, S. J. (2007). Rational numbers and proportional reasoning: toward a theoretical framework for research. In F. K. Lester Jr. (ed.), *Second handbook of research on mathematics teaching and learning* (pp. 629–68). Charlotte, NC: Information Age Publishing.

Lexico. (2020). Estimate. Retrieved from https://www.lexico.com/definition/estimate

Livy, S. & Herbert, S. (2013). Second-year pre-service teachers' responses to proportional reasoning test items. *Australian Journal of Teacher Education*, 38(11). Retrieved from http://ro.ecu.edu.au/ajte/vol38/iss11/2

Lyons, I. M. & Beilock, S. L. (2012). When math hurts: math anxiety predicts pain network activation in anticipation of doing math. *PLoS ONE*, 7(10), e48076.

Middleton, J. A. (2013). More than motivation: the combined effects of critical motivational variables on middle school mathematics achievement. *Middle Grades Research Journal*, 8(1), 77–95.

Moloney, K. & Stacey, K. (1997). Changes with age in students' conceptions of decimal notation. *Mathematics Education Research Journal*, 9(5), 25–38.

Olson, A. M. & Stoehr, K. J. (2019). From numbers to narratives: preservice teachers' experiences with mathematics anxiety and mathematics teaching anxiety. *School Science and Mathematics*, 119, 72–82.

Parkes, A. A., Couchman, K. E., Jones, S. B. & Green, K. N. (1967). *Betty and Jim: Year 6 mathematics*. Sydney: Shakespeare Head Press.

Pendergast, D. (2017). School reform and sustainable practice. In D. Pendergast, K. Main & N. Bahr (eds.). *Teaching middle years: rethinking curriculum, pedagogy and assessment* (pp. 323–58). Crows Nest, Australia: Allen and Unwin.

Polya, G. (1957). *How to solve it* (2nd edn). Princeton, NJ: Princeton University Press.

Population Education. (2016). Food for thought. Retrieved www.lexico.com/definition/estimatewww.populationeducation.org/sites/default/files/food_for_thought.pdf

Quinnell, R., Thompson, R. & LeBard, R. J. (2013). It's not maths; it's science: exploring thinking dispositions, learning thresholds and mindfulness in science learning. *International Journal of Mathematical Education in Science and Technology*, 44(6), 808–16.

Ramirez, G., Yang Hooper, S., Kersting, N. B., Ferguson, R. & Yeager, D. (2018). Teacher math anxiety relates to adolescent students' math achievement. *AERA Open*, 4(1), 1–13.

Reys, R. E., Lindquist, M. M., Lambdin, D. V., Smith, N. L., Rogers, A., Cooke, A., Ewing, B., Robson, K. & Bennett, S. (2017). *Helping children learn mathematics* (2nd edn). Milton, Qld: Wiley.

Sellars, M. (2018). Mathematics and numeracy in a global society. In M. Sellars (ed.). *Numeracy in authentic contexts* (pp. 5–21). Singapore: Springer Nature.

Siemon, D., Beswick, K., Brady, K., Clark, J., Faragher, R. & Warren, E. (2011). *Teaching mathematics: foundations to middle years*. South Melbourne: Oxford University Press.

Sowder, J., Armstrong, B., Lamon, S. J., Simon, M., Sowder, L. & Thompson, A. (1998). Educating teachers to teach multiplicative structures in the middle grades. *Journal of Mathematics Teacher Education*, 1, 127–55.

Steen, L. A. (1998). Numeracy: the new literacy for a data-drenched society. *Educational Leadership*, 57(2), 8–13.

Sullivan, P., Bobis, J., Downton, A., Hughes, S., Livy, S., McCormick, M. & Russo, J. (2019). Ways that relentless consistency and task variation contribute to teacher and student mathematics learning. In A. Coles (ed.). *For the Learning of Mathematics Monograph 1: proceedings of a symposium on learning in honour of Laurinda Brown* (pp. 32–7). Canada: FLM Publishing Association.

Tatto, M. T., Peck, R., Schwille, J., Bankov, K., Senk, S. L., Rodriguez, M., Ingvarson, L., Reckase, M. & Rowley, G. (2012). *Policy, practice, and readiness to teach primary and secondary mathematics in 17 countries: findings from the IEA Teacher Education and Development Study in Mathematics (TEDSM)*. Amsterdam: International Association for the Evaluation of Educational Achievement (IEA).

Tosto, M. G., Petrill, S. A., Malykh, S., Malki, K., Haworth, C. M. A., Mazzocco, M. M. M., Thompson, L., Opfer, J., Bogdanova, O. Y. & Kovas, Y. (2017). Number sense and mathematics: which, when, and how? *Developmental Psychology*, 53(10), 1924–39.

Van De Walle, J. A., Karp, K. S. & Bay-Williams, J. M. (2014). *Elementary and middle school mathematics: teaching developmentally* (8th edn). Harlow, UK: Pearson.

Way, J. (2011). Number sense series: developing early number sense. Retrieved from https://nrich.maths.org/2477

Wilkie, K. J. & Sullivan, P. (2018). Exploring intrinsic and extrinsic motivational aspects of middle school students' aspirations or their mathematics learning. *Educational Studies in Mathematics*, 97, 235–54.

# INDEX